LIST OF TITLES

Already published

Editors' Foreword

The student of biological science in his final years as an undergraduate and his first years as a graduate is expected to gain some familiarity with current research at the frontiers of his discipline. New research work is published in a perplexing diversity of publications and is inevitably concerned with the minutiae of the subject. The sheer number of research journals and papers also causes confusion and difficulties of assimilation. Review articles usually presuppose a background knowledge of the field and are inevitably rather restricted in scope. There is thus a need for short but authoritative introductions to those areas of modern biological research which are either not dealt with in standard introductory textbooks or are not dealt with in sufficient detail to enable the student to go on from them to read scholarly reviews with profit. This series of books is designed to satisfy this need. The authors have been asked to produce a brief outline of their subject assuming that their readers will have read and remembered much of a standard introductory textbook of biology. This outline then sets out to provide by building on this basis, the conceptual framework within which modern research work is progressing and aims to give the reader an indication of the problems, both conceptual and practical, which must be overcome if progress is to be maintained. We hope that students will go on to read the more detailed reviews and articles to which reference is made with a greater insight and understanding of how they fit into the overall scheme of modern research effort and may thus be helped to choose where to make their own contribution to this effort. These books are guidebooks, not textbooks. Modern research pays scant regard for the academic divisions into which biological teaching and introductory textbooks must, to a certain extent, be divided. We have thus concentrated in this series on providing guides to those areas which fall between, or which involve, several different academic disciplines. It is here that the gap between the textbook and the research paper is widest and where the need for guidance is greatest. In so doing we hope to have extended or supplemented but not supplanted main texts, and to have given students assistance in seeing how modern biological research is progressing, while at the same time providing a foundation for self help in the achievement of successful examination results.

General Editors:

W.J. Brammar, Professor of Biochemistry, University of Leicester, UK

M. Edidin, Professor of Biology, Johns Hopkins University, Baltimore, USA

The Cytoskeleton:
Cellular architecture and choreography

Alice Fulton

Department of Biochemistry
University of Iowa

New York   London
Chapman and Hall

First published 1984 by
Chapman and Hall
733 Third Avenue, New York NY 10017
Published in the UK by
Chapman and Hall Limited
11 New Fetter Lane, London EC4P 4EE

© 1984 A. Fulton

Printed in Great Britain by
J.W. Arrowsmith Ltd, Bristol

ISBN 0 412 25510 3

British Library Cataloguing in Publication Data

Fulton, Alice
    The cytoskeleton.
    1. Plant cell walls
    I. Title
    581.87'34   QK725

    ISBN 0-412-25510-3

Library of Congress Cataloging in Publication Data

Fulton, Alice
    The cytoskeleton.

    (Outline studies in biology)
    Bibliography: p.
    Includes index.
    1. Cytoskeleton. I. Title. II. Series: Outline
studies in biology (Chapman and Hall)
QH603.C96F84 1984   574.4'7   84-9483
ISBN 0-412-25510-3

Contents

Acknowledgements

To the many colleagues whose generous help made this book better.
To Sheldon Penman, for the use of micrographs, and to many whose
    kind offers were declined only for lack of space.
To Carol Galbraith, who made writing this book as easy as it could be.
To Tom, who made this book possible.

My thanks.

1 Introduction

In eucaryotic cells, Brownian motion is a sign of death. The living cell orients its organelles and subcellular particles, so that everything not at rest undergoes directed and metabolically driven motion. Moreover, the patterns of motion and cell shape are specific; every cell type can be characterized by a particular configuration. This ceaseless and pervasive organization is mediated through the cytoskeleton; the cytoskeleton can be functionally defined as the structures responsible for this spatial organization.

What is the cytoskeleton? For some, the cytoskeleton is solely the network of intermediate filaments, or of microfilaments, or of microtubules. For others, it is a combination of two or three of these systems of filaments and their associated proteins. It is unlikely, however, that any one of these filament systems has an exclusive claim to being the cytoskeleton, since these three filament systems interact extensively.

The preceding definitions of the cytoskeleton are optative. An operational approach to the structural basis for cellular shape and location is to define the cytoskeletal framework operationally as the structures that remain after a non-ionic detergent extraction. The cytoskeletal framework includes the aforementioned filament systems when the extraction is performed appropriately. However, the structure obtained is complex and preserves many relationships to the nuclear matrix and cell membrane not included in the narrower definitions above. A yet more inclusive term for the structures responsible for spatial organization is the cytoplasmic matrix, visible when intact cells are fixed. It probably includes proteins whose associations with the cytoskeletal framework are briefer than the cytoskeletal filaments in the stricter sense. The cytoplasmic matrix includes many small heterogeneous elements termed microtrabeculae, that interconnect elements of the cytoskeleton, the cytoskeletal framework and subcellular organelles.

These definitions shade into each other; any attempt to be overstrict leads to difficulties such as those by which peripheral membrane proteins would be excluded from membranes proper. What should be clear from this discussion is the hierarchical nature of the cytoskeleton. The mechanisms whereby the cell is spatially organized lie in the chemical properties of the cytoskeletal proteins. These proteins are integrated in a tissue-specific and cell-specific manner into complex, dynamic networks and structures that interact with each other. These structures,

7

in their turn, are responsible for the co-ordinated spatial behavior of the cell. In the end, a complete explanation of the spatial behavior of the cell must be found in the molecules responsible for it. Likewise, our understanding of the cytoskeleton will not be complete until we can account for the full and complex choreography of the cell.

Additional reading

[1] Borgers, M. and De Brabander, M. (1975), *Microtubules and Microtubule Inhititors*, North-Holland Publishing Co., Amsterdam.
[2] Cappuccinelli, P. and Morris, N.R. (1982), *Microtubles in Microorganisms*, Marcel Dekker, Inc., New York and Basel.
[3] Ciba Foundation Symposium (1973), *Locomotion of Tissue Cells*, Associated Scientific Publishers, Amsterdam.
[4] Cold Spring Harbor Symposia on Quantitative Biology, XLVI (1982), *Organization of the Cytoplasm*, Cold Spring Harbor Labs.
[5] Curtis, A.S.G. and Pitts, J.D. (1980), *Cell Adhesion and Motility*, Cambridge University Press, Cambridge.
[6] Dowben, R.M. and Shay, J.W. (1981), *Cell and Muscle Motility*, Plenum Press, New York.
[7] Dustin, P. (1978), *Microtubles*, Springer-Verlag, New York.
[8] Frederiksen, D.W. and Cunningham, L.W. (1982), *Methods in Enzymology, Structural and Contractile Proteins; Part B, The Contractile Apparatus and the Cytoskeleton*, Academic Press, New York.
[9] Goldman, R., Pollard, T., Rosenbaum, J. (1976), *Cell Motility*, Cold Spring Harbor.
[10] Pearson, M.L. and Epstein, H.F. (1982), *Muscle Development: Molecular and Cellular Control*, Cold Spring Harbor.
[11] Roberts, K. and Hyams, J.S. (1979), *Microtubules*, Academic Press, London.
[12] Tanenbaum, S.W. (1978), *Cytochalasins: Biochemical and Cell Biological Aspects*, North-Holland Publishing Co., Amsterdam.

2 Protein chemistry

The proteins that compose the filaments of the three major cytoskeletal networks have been isolated and their behavior *in vitro* is beginning to be well known. Actin, tubulin and the intermediate filament proteins can all be disassociated and reassembled *in vitro* to permit analysis of their chemical properties. Several common features have emerged which reflect the functional constraints and interactions of these systems. All three proteins form filaments, an economical structural member. These filaments are polar; their heads and tails are chemically distinct. In addition, these filaments are helical. Helical structures possess multiple equivalent binding sites; thus, they can form a multiplicity of structures generated by probabilistic interactions rather than the deterministic pathways followed during phage assembly [1]. These proteins all exist in multiple forms. Thus, there are several tissue-specific isoforms of actin and tubulin. Nonetheless, they are evolutionarily highly conserved [2]. For the intermediate filament proteins, tissue-specific variation is more conspicuous and evolutionary conservation less so [3, 4]. Nonetheless, antibodies exist that recognize intermediate filaments from *Drosophila* and mammals. Calcium plays a role in the regulation and function of all three filament systems and enters into the behavior of the cytoskeleton in so many highly specific and regulatory ways, that calcium might almost be regarded as the cell's transformer of soluble signals into solid cytoskeletal configurations. All three of these filament-forming proteins can associate with other proteins, either in the filament form or otherwise, and it is these specific associations that will be discussed at length in this chapter. Phosphorylation plays many roles in these interactions, either through the phosphorylation of associated proteins or of the filament-forming protein itself. Finally, the cytoskeletal filament proteins, are conspicuous by their acidity. The amino-terminus of actin, the carboxy-terminus of tubulin and the intermediate filament proteins in general are highly acidic and the role that this acidity plays is still being uncovered.

2.1 Actin and actin-binding proteins

2.1.1 *Actin polymerization*
Actin monomer (G-actin, 42 kD) is a globular protein with binding sites for divalent cations and nucleotides. Under physiological conditions

these binding sites are occupied by magnesium and ATP. Actin polymerization is a multistep process; the influence of actin-binding proteins can only be understood if the full complexity of the process is kept in mind [5]. Polymerization is represented diagrammatically in Fig. 2.1, and results in filaments (F-actin).

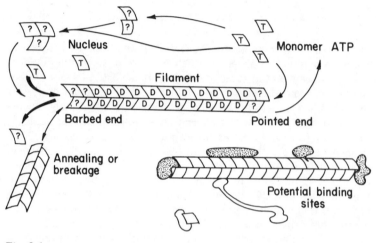

Fig. 2.1

Actin monomers can associate to form both dimers and trimers, but only the trimer is sufficiently stable to nucleate further polymerization. Once a nucleus is formed (probably trimeric), elongation can proceed rapidly. Elongation from a nucleus can proceed from both ends of the filament, but under most conditions the rates of elongation will not be identical. The polarity of the filament can be revealed by decoration with heavy meromyosin, which produces an arrow-headed pattern on the filament. The barbed end of the filament after myosin decoration is also the fast assembly end and the end of the filament with the lower critical concentration in the presence of ATP. The pointed end is the slower growing end and has a higher critical concentration. Polymerization is associated with ATP hydrolysis, but does not require it since polymerization goes equally well in the presence of non-hydrolyzable analogs [5]. Finally, as filaments elongate, thermal agitation may lead to breakage, and filaments may anneal with other filaments.

Thus, there are conceptually several distinct processes in actin polymerization: nucleation, addition at either end, dissociation from either end, filament breakage and annealing. Unfortunately, in a solution of actin induced to polymerize, most of these processes will be occurring simultaneously. In addition, each time filaments break they generate new free ends that behave like nuclei. Thus, detailed analysis of the separate processes requires intervention in some form to discriminate between the different aspects of polymerization.

The method of measuring polymerization also constrains which aspect of polymerization will be visible. For example, viscosity is sensitive to filament length, but not to subunit exchange. Fluorescence quenching measurements are insensitive to filament length, and thus good measures of fractional incorporation into filaments, but they are sensitive to the presence or absence of subunit exchange [5]. Direct measurements of filament length in the electron microscope, unless limited to short filaments, encounter the difficulty of filament breakage. Thus, the complete characterization of the kinetic and equilibrium properties of actin polymerization of actin-binding proteins usually requires several different methods of measurement.

2.1.2 *Actin-binding proteins*
From the discussion above, it is clear there are five principal sites at which actin-binding proteins may have their effect. They may bind to the monomer, the pointed or slow assembly end of the filament, the barbed or fast assembly end of the filament, the side of the filament, or between two filaments to serve as crosslinkers (Fig. 2.1). In addition to these five modes of interaction, such proteins may be sensitive or insensitive to calcium. With so many varieties of interaction it is hardly surprising that a multiplicity of actin-binding proteins have been described, some with several modes of interaction.

Monomer-binding proteins inhibit nucleation by reducing the interaction of monomers. They may or may not lower the rate of elongation, depending upon whether or not the monomer associated with the actin-binding protein can associate with filaments. Two calcium-sensitive monomer-binding proteins are profilin and fragmin [6]. Both require calcium to bind to the actin monomer. The profilin–monomer complex can however polymerize onto existing filaments. Fragmin–actin complexes will not. Thus, profilin principally suppresses nucleation whereas fragmin both suppresses nucleation and inhibits the rate of elongation. Of the three calcium-insensitive monomer-binding proteins, DNase I [5] and vitamin D-binding protein [5] are both extracellular. The significance of their ability to bind actin is not known. There is an actin-depolymerizing protein from brain, however, which depolymerizes filaments by binding monomer and thus lowering the available concentration of actin [5].

The barbed or fast assembly end of actin filaments can be blocked either by capping proteins or the drug cytochalasin B or D. By blocking the fast assembly end, capping proteins promote nucleation but inhibit elongation and annealing, and their overall effect is often to shorten filaments, both by creating more nuclei that compete for free monomer and by preventing annealing. At least four proteins are known to have these effects in the presence of calcium. Gelsolin [5], a 90 kD protein from platelets [5], villin [5], and fragmin [5] can all eliminate the lag during nucleation for pure monomer and also can shorten existing filaments in the presence of calcium. Calcium-insensitive capping proteins also exist. Acanthameba 31 kD and 28 kD proteins and a platelet 65 kD protein [6] have similar effects, whether or not calcium is present.

The pointed or slow assembly end of the filament is another potential site for protein interaction. A protein binding here could nucleate assembly and prevent annealing. It would affect the elongation rate of actin as a function of actin concentration; between the critical concentrations of the slow assembly end and the fast assembly end, a protein binding to the slow end would increase elongation by preventing loss of monomers from the pointed end. However, above the higher critical concentration, the net effect of a protein binding to the slow end would be to lower the overall elongation rate by blocking one of the two sites of addition. The net effect of these three properties (increased nucleation, decreased annealing and elongation) is to increase the number and shorten the length of filaments. These effects are similar to those of proteins that bind at the barbed end. Therefore, distinguishing between these two classes requires either determining a protein's effect in competition with proteins known to bind to the fast end, or using a seeded assembly assay [6, 5] to determine which end the protein affects. So far, only one protein is known that definitively binds to the pointed or slow assembly end of actin filaments, namely acumentin, an abundant protein from macrophages [7]. Another possible candidate is brevin [6], a serum protein which rapidly decreases the viscosity of F-actin solutions by shortening the filaments without increasing the free monomer concentration. Both brevin and acumentin are insensitive to the calcium concentration.

A fourth potential binding site is along the side of the actin filament without further interactions with other filaments. Proteins binding at the side of the filament might then either stabilize the filament or destabilize it. Tropomyosin binds in a calcium-insensitive way to stabilize F-actin filaments [5], while severin and villin bind to F-actin filaments and cut them in the presence of calcium [5, 8].

Probably the most dramatic of the actin-binding proteins are those which can crosslink actin filaments and form gels. By virtue of their binding to the F-actin form, such proteins generally nucleate as well as gel actins. At least four crosslinking proteins will cause F-actin to gel in the absence of calcium. Platelet α actinin [5], villin [6], fimbrin [9], and macrophage actinogelin [6] will all crosslink an F-actin solution into a rigid gel that will not permit the passage of a metal ball. The addition of calcium will dissolve this gel. All four of these proteins are monomeric. In the case of villin, the protein can be dissected into separate domains, a core which is calcium sensitive and can bind to and cap actin filaments and a head region that is required for crosslinking filaments in the absence of calcium. There are also numerous calcium-insensitive crosslinking proteins. Two of these, filamin and actin binding protein [6], originally found in macrophages, share the properties of being a homodimer of long, flexible proteins. Alpha-actinin from muscle is also a calcium-insensitive crosslinking protein [5]. Vinculin and a high molecular weight protein from BHK cells are also capable of crosslinking without additional proteins. However, fascin from sea urchins can only generate actin needles by itself. It requires the co-operation of a 220 kD protein to generate gels [6].

The spectrin family of proteins is one of the most interesting of the crosslinking proteins that are not directly affected by calcium. Spectrin is an $(\alpha\beta)_2$ tetramer first found in the membrane cytoskeleton of red blood cells. The $\alpha\beta$ dimers are associated tail-to-tail, leaving the heads of the molecule free to interact with actin oligomers. In addition, the α subunit of each dimer can interact with calmodulin, a calcium-binding protein involved in many calcium regulated activities. It is not known as yet what effect calmodulin binding has on the activity of spectrin. Spectrin-like molecules have now been found in a wide variety of cell types [10,11,12], so that it is more appropriate to speak of the spectrin family. The α subunit of red blood cell spectrin is 240 kD, and a 240 kD immunologically cross-reactive protein has been found in most cell types examined. However, the β subunit of red blood cell spectrin is 220 kD. Cells that display a 240 kD protein that reacts with α spectrin antibodies may also display, for example, a 260 kD subunit (found in the terminal web) or a 235 kD subunit (found in nerve cells and elsewhere). These cross-reacting species were originally described as separate families, called TW260/240 and fodrin. However, spectrin appears to be a tissue-specific family, as has been seen for so many cytoskeletal proteins. That all of these proteins contain a calmodulin-binding domain is a recent discovery and the consequences of that activity are yet to be understood.

Myosin is the only actin-associated protein that can generate mechanical force. This ATP derived mechanical work is the basis for muscle contraction and is believed to generate the tension exerted by fibroblasts and other cells in contact with the extracellular matrix. The interaction of myosin and actin is complex, so much so that it has been discussed in another book of this series. Myosin performs work by a cyclical interaction with actin. Myosin—ADP can bind actin filaments. A change in conformation occurs that is accompanied by ADP release. If ATP is available in solution to replace the ADP released from the myosin, it causes the release of the actin filament; ATP hydrolysis permits the initiation of a second cycle. Calcium regulates this interaction at any of several sites. In some muscle cells, calcium interacts with troponin to control the binding of tropomyosin to actin. These cells

13

are considered to be regulated at the thin filament. In other muscles, calcium acts on the myosin molecule, either directly or by activating enzymes that phosphorylate the light chains of the myosin molecule. In some non-muscle cells, calcium regulates contraction at the level of assembly of the myosin filaments.

The relationship between these different classes of actin-binding proteins gains some focus when considered in terms proposed by Florey's gel theory. Gel theory predicts that, when polymers have a greater than even probability of connecting to other polymers in the system, a three-dimensionally connected network will result. This predicts an abrupt transition from solution to gel, called the gel point, which has some mathematical similarities to other transitions, such as melting or vaporization. Beyond the gel point, further crosslinking leads to differences in rigidity only. Thus, crosslinking proteins can transform a viscous solution of F-actin into a solid gel. Filament-breaking proteins or proteins which increase the filament number can dissolve this gel by decreasing the average length of polymer without increasing the number of crosslinks. When the necessary density of crosslinks falls below that required for the gel point, the actin gel will dissolve. Myosin can interact with a gel to contract it. Gel theory is helpful in connecting the properties of the different classes of actin-binding proteins, and it has proven valuable for developing assays for different functions. It is probably important to bear in mind, however, that gel theory discusses isotropic structures and does not concern itself with precise topological connections. As will be clear later, these topological properties are extremely important properties of the cytoskeleton that cannot yet be predicted by gel theory.

Interpreting protein chemistry in a meaningful way requires detailed knowledge of the conditions that obtain within the cell. This includes the precise stoichiometry of all the relevant proteins and determinations of other controlling conditions such as pH, pCa nucleotide concentrations, and probably other variables such as phospholipid composition of adjoining membranes. When proteins can be effective at stoichiometries of 1:500 in promoting phenomena that exhibit cooperative and abrupt transitions, quantitative prediction will clearly be a challenging task.

2.2 Tubulin and microtubule polymerization

Microtubules, like microfilaments, are linear polymers, formed from tubulin subunits. However, the tubulin subunit is an $\alpha\beta$ dimer; both α- and β-tubulin are approximately 55 kD and can bind GTP or GDP. In the dimer, however, only nucleotide bound to β tubulin can exchange with GTP in solution. Like actin, tubulin has a highly conserved protein sequence. The α and β peptides diverged early in eucaryotic evolution; subsequent change has been less extensive [13].

Tubulin polymerization shares many characteristics with actin polymerization. The subunits must associate to form a nucleus from which there is biased bi-directional growth with associated hydrolysis of the bound nucleotide triphosphate [13]. As a consequence, the two ends of the microtubule have different critical concentrations and permit potentially the treadmilling of subunits through the microtubule when the free subunit concentration lies between the critical concentrations of the two ends [14]. Again, as for actin filaments, breakage and annealing of the microtubules can change the number density of microtubule ends without changing the number of subunits found in the polymer.

Microtubule polymerization is affected by the concentration of divalent cations and temperature; i.e., it is inhibited by calcium, EDTA, and cold. Hydrolysis of GTP is not required for polymerization since nonhydrolyzable analogs support polymerization at normal rates. The microtubules so formed are now stable to calcium.

Microtubule polymerization, however, has a larger number of potential pathways than actin polymerization. *In vitro* several polymorphs have been observed, depending upon the conditions of polymerization [13]. These polymorphs led to efforts to find factors that promote polymerization, often under rather unphysiological conditions. In principle, tubulin-binding proteins could be classified as we have classified actin-binding proteins, that is, as binding to the subunit, the fast assembly or the slow assembly end, or the side of the filament. However, for historical reasons, the majority of microtubule-associated proteins have been studied either in terms of co-polymerization with microtubules, or of their ability to stimulate assembly. From the discussion above for microfilaments, it is clear that 'assembly' is the sum of nucleation, elongation, breakage and annealing, and that each of these steps can be affected by proteins that bind to tubulin. In addition, since the nucleus required for microtubule elongation is larger than that required for microfilaments, nucleation is especially sensitive to tubulin concentration. Any factor which stabilizes nuclei will be most conspicuous in its enhancement of nucleation, whether or not that is its function or site of association *in vivo*. With these caveats, the microtubule-associated proteins will be discussed individually below.

2.2.1 *Tubulin associated proteins*

Two major groups of microtubule-associated proteins were originally identified either by their co-purification with tubulin during cycles of assembly and disassembly or by their association with tubulin during purification by other means. A persistent association with tubulin is not an adequate criterion for specific association, since tubulin is a highly charged protein and can behave as an ion exchange resin. However, these two major groups have also been found on microtubules in cells fixed without extraction.

One group of microtubule-associated proteins (MAPs) is the group of high molecular weight (HMW) MAPs, between 290 and 350 kD in weight, which are especially common in brain microtubules [15]. A second group of microtubule associated proteins, called tau, are between 55 and 70 kD [16, 17]. Although both HMW and tau were primarily described as promoting polymerization, the appearance of microtubules decorated with MAPs from either group and the location of these proteins in the cell make it clear that both HMW and tau associate with the side of microtubules. Their enhancement of polymerization appears to be by promoting nucleation, presumably by a mechanism similar to the side-binding proteins that associate with actin. In addition, microtubules decorated with HMWs display side arms that give the microtubules a furry appearance. These side arms are capable of associating with secretory granules; this association is reversed by ATP [18].

A different method of identifying microtubule-associated proteins, which exploits associations with microtubules *in vivo*, has identified a number of MAP proteins in various cells [19]. These include a 69 kD protein that is homologous to the tau proteins, and an 80 kD MAP protein with peptide homologies to the 69 kD protein. Both the 69 kD and 80 kD proteins are phosphorylated to various extents; the more highly phosphorylated forms are more extensively associated with microtubules [19]. This is one of the few cases in which the phosphorylation of a cytoskeletal protein has a measurable effect on its affinity for the cytoskeleton.

Nucleoside diphosphate kinase has been found associated with microtubules and can associate at a constant specific activity through three cycles of polymerization [13]. It is not identical to either of the HMW proteins or to tau. It is capable of phosphorylating both GDP and ADP. However, its contribution to microtubule function is not yet known.

Tubulin L-tyrosine ligase is an enzyme found in two isoforms. Tyrosine ligase catalyzes the addition of a tyrosine post-translationally to the C-terminal glutamate of the α tubulin protein. Variations in the distribution of both the enzyme and of the tyrosinylated tubulin have been noted, but the functional significance is still unknown [13].

Cyclic AMP-dependent protein kinase has been found to co-purify with microtubules purified by several different techniques. This kinase is associated with MAP 2 and will phosphorylate both MAP 2 and tau in the presence of cyclic AMP and ATP [20]. In light of the increased association of phosphorylated tau proteins with microtubules, it seems likely that such phosphorylation might increase the stabilizing effect of these proteins on microtubules or increase the extent of association of these proteins to microtubules.

The high molecular weight MAPs, MAP 1 (approximately 350 kD) and MAP 2 (approximately 270 kD), are both heterogeneous on SDS PAGE gels. MAP 2 has been separated into two species, 2A and 2B. MAP 1 consists of at least three species, MAP 1A, 1B, and 1C. MAP 1 possesses smaller polypeptides, called light chains 1 and 2, approximately 30 and 28 kD, which appear to be present in a 1:1 stoichiometry.

MAP 1 appears to be more widely distributed than MAP 2 and has been seen both on interphase microtubules and on the mitotic spindle [21].

Microtubules can associate with a mechanochemical protein, dynein, which is in many respects analogous to myosin and its interactions with microfilaments [22]. The composition of dynein depends upon its source and the methods used to isolate it. Ciliary dynein consits of approximately 12 polypeptide chains, including three of ~400 kD, two or three of ~85 kD, and six to eight small chains. Dynein exhibits a microtubule-stimulated ATPase which is inhibited by orthovanadate. Various treatments retain the ATPase activity of dynein but eliminate its ability to crosslink microtubules. This suggests that there are several functional domains for binding, just as on many of the actin crosslinking proteins. Dynein can crosslink both doublet microtubules from cilia and singlet microtubules formed by purified brain tubulin. These crosslinks are disaggregated by ATP. Dynein binds to microtubules in a polar orientation, and microtubules can be decorated with purified dynein to display their intrinsic polarity.

Arguing by analogy to actin, it is conceivable that both monomer-binding proteins and end-capping proteins also exist for tubulin. However, no directed search has been undertaken for such proteins. It is fairly straightforward to predict the properties of such proteins if obtained, but their existence at present is speculative. The presence within the cell of nucleating centers (v. i.) suggests the existence of as yet unidentified end-binding proteins.

Microtubules within the cell do not form gels as actin filaments appear to. Therefore, the principal properties of tubulin-binding proteins of interest are ones which stabilize microtubules and which cross link microtubules, either to other microtubules, intermediate filaments, or subcellular particles such as secretory granules. The characterization of such proteins is still in its infancy with the exception of dynein and MAP 2.

2.3 Intermediate filament proteins

The third major class of filaments for which we can definitively assign a structural protein are the intermediate filaments. These filaments are approximately 10 nm in diameter and have been found in most vertebrate cells. The correct assignment of the structural proteins has been more recent for these filaments; as a consequence, less is known about the associated proteins. In addition, the intermediate filaments display a much larger range of tissue-specific variation in their sequence. Thus there was some controversy during the assignment of the different intermediate filament proteins to their respective filaments.

All intermediate filaments share certain properties. They are helical, which confers a periodic ultrastructure to intermediate filaments under the appropriate staining conditions. All of them have been disassembled and reassembled *in vitro*, but for all of them, the *in vitro* assembly conditions are such that disassembly and reassembly in the cell are likely to be rare events [23], in contrast to actin and tubulin. Finally, a common structural model has been proposed for intermediate filaments. Initially it was proposed that the filaments consist of a triple chain coiled coil and that the coiled coil regions were interrupted by non-helical domains at the amino-terminus, the carboxy-terminus, and an intervening non-helical domain [24]. More recent studies, based on the amino acid sequence for several of these proteins, have suggested that intermediate filaments are a two-fold coiled coil with the coiled coil domains spaced somewhat differently from that suggested by the proteolytic studies [25]. A significant point to emerge from both these studies is that the non-coiled regions are very sensitive to proteolysis; proteolysis of some of these domains leads to the disassembly of the filaments.

The intermediate filament proteins (IFP) can be organized into five families. Because of minor differences between species and apparent molecular weight under different analysis conditions, these families have numerous synonyms. Epithelial cells generally contain cytokeratins, a family of keratin-like proteins between 45 and 60 kD molecular weight. Neuronal cells contain neural filament proteins with major polypeptides of 68 kD, 145 kD, and 220 kD. Muscle cells contain desmin or skeletin of approximately 53 kD. Glial cells contain glial fibrillary acidic protein (GFAP) of approximately 55 kD and mesenchymal cells contain vimentin or decamin of approximately 58 kD. The apparent molecular weights of these proteins vary somewhat with the source of the filaments and with the gel electrophoresis system used to analyze them.

All of these proteins have been disassembled and reassembled *in vitro*; desmin, vimentin, and GFAP each reform as a homopolymer of a single polypeptide. The neuronal IFP can be separated into three distinct polypeptides. Under appropriate conditions, the 68 kD protein will reassemble into a homopolymer in the absence of the 145 kD and 220 kD proteins. However, the resulting neurofilaments are smooth-walled rather than having the slightly fuzzy appearance of the native filaments [23]. This suggests that the 220 kD protein and possibly the 145 kD protein are partially peripheral in the native filament, and confer upon the filament the side arms visible in the electron microscope. These side arms are especially attractive as potential sites of interaction with other proteins or with subcellular particles.

The cytokeratins include multiple polypeptides. Epithelia from different parts of the body each contain a different spectrum of cytokeratins [26]. To date, it has been impossible to assemble *in vitro* an intermediate filament out of a single cytokeratin, although various combinations of two or more can reassemble *in vitro*. This limited reassembly potential of the cytokeratins has made it possible to demonstrate the co-assembly of cytokeratins with other intermediate filament proteins. By choosing conditions under which neither intermediate filament protein alone would form filaments, it has been possible to demonstrate the assembly of mixtures of heterologous intermediate filament proteins [27].

Amino acid sequences show that the intermediate filament proteins are related both between different species for a given tissue type and between the different tissue forms of intermediate filament proteins. However, the intermediate filament proteins from the same tissue in two different species will resemble each other more than intermediate filaments from two different tissues. For example, vimentin and desmin from pig have 64% sequence identity whereas desmins taken from pig and chicken show 91% identity [28]. Thus, even though the intermediate filament proteins show different isoelectric behavior and different apparent molecular weights, all five groups are clearly related ancestrally. This sequence relatedness is in turn expressed in the common ultrastructure of the filaments.

This pattern of homology and differentiation has made it possible to generate antibodies that either recognize specific tissue forms, e.g., vimentin from many different animals, or that recognize domains common to many intermediate filament proteins. One of these broad specificity antibodies has been used to indicate the presence of intermediate filament proteins in *Drosophila* cells, again suggesting that this family of proteins is at least as ancient as the metazoa [29].

The intermediate filament proteins are chemically diverse and generally insoluble. As a consequence, few intermediate filament associated proteins have so far been analyzed. One of the most interesting is a calcium-activated protease that is specific to intermediate filament proteins. Such calcium-activated proteases have been observed associated with vimentin [30], neurofilaments [31], and desmin [32]. The presence of the protease on the intermediate filament ensures a close association between the substrate and the protease. The optimum concentration of calcium for activation is 10 μM for the vimentin-specific protease. Proteolysis rapidly disassembles the filaments. This protease is particularly intriguing since the stability of intermediate filaments *in vitro* is such that they would not be expected to extensively assemble or disassemble under physiological conditions, and degradation may be the only route available to the cell for disassembling filaments.

Only a few IFP-associated proteins have been characterized. Plectin is a 300 kD protein isolated from the intermediate filaments of glioma cells [33]. In these cells, plectin is found associated with vimentin in approximately a 1:20 stoichiometry. Although the plectin shows some co-localization with intermediate filaments it is not present over the entire length of these filaments. *In vitro*, plectin can bundle vimentin filaments, indicating that it either contains two vimentin-binding domains per 300 kD polypeptide or that the native protein is a dimer or tetramer.

Synemin is a 230 kD protein that associates with both desmin and vimentin filaments [34]. Synemin has been isolated from muscle cells and from avian erythrocytes. Paranemin (280 kD) associates with both desmin and vimentin filaments in avian skeletal muscle. It is particularly conspicuous early in development and is reduced at later stages, suggesting a transient function for paranemin [35]. Paranemin is found associated both with vimentin and desmin filaments, whether or not those filaments also include synemin.

2.4 Proteins that associate with several filament systems

The filament-associated proteins have been discussed so far as if each has a unique specificity. This approach has been used in many biochemical studies and has proven fruitful for isolation and characterization. However, both ultrastructurally and behaviorally, the filament systems are not independent, and crosslinks are seen between them. If nothing else, this would suggest that there are proteins that could potentially be identified by their ability to crosslink two classes of filaments. It is, therefore, particularly exciting that recent studies have shown that some reasonably well-characterized proteins in fact are multifunctional. Some of these multiple relationships are shown in Fig. 2.2. MAP 2, a high molecular weight microtubule-associated protein which associates with cyclic AMP-dependent protein kinase, also interacts with neuronal intermediate filaments [36] and with actin. Perhaps most striking is that these interactions with other filament systems can be regulated by ATP. That is, in the presence of ATP, MAP 2 causes an increase of viscosity in mixtures of microtubules and

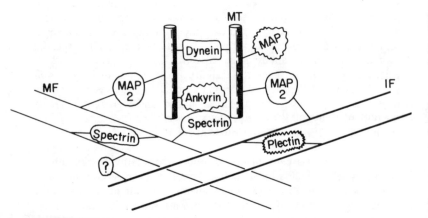

Fig. 2.2

neurofilaments. In contrast, MAP 2 interacts with actin in the absence of ATP if cAMP is present. Both the intermediate filament and microfilament interactions are calcium insensitive [22]. It is perhaps not surprising that MAP 2 has substantial homologies to spectrin, although some domains of each protein are clearly unique.

Ankyrin is a second bifunctional protein. Ankyrin binds to spectrin, an actin-binding protein. More recently, ankyrin has also been shown to bind to microtubules; purified red blood cell ankyrin will bind to cycle-purified brain microtubules [37]. Antibodies to ankyrin reveal that it has a degree of homology to MAP 1. MAP 1 is also in the molecular weight range of plectin, the intermediate filament bundling protein, and a possible homology between MAP 1 and plectin has not yet been ruled out. The only two filament systems not yet known to be linked by bifunctional proteins are the microfilaments and intermediate filaments. MAP 2 cannot serve this function in the presence of cAMP, since it interacts with microfilaments in the absence of ATP and with intermediate filaments in its presence. It is possible that these two filament systems are linked only through the microtubules or at low levels of cAMP. It would be unduly sanguine to hope that the cell's repertoire of cytoskeletal proteins is exhausted; it is quite likely that proteins capable of interacting with two or more filament systems will continue to be discovered.

A different kind of bifunctionality is displayed by proteins which engage the filament systems and various membrane components of the cell. Tubulin has been found associated with clathrin-coated vesicles [38], suggesting that at least some of these vesicles move over microtubules. In addition, at least two groups of proteins allow the microfilament system to interact with the cell membrane. Microfilaments associate with the membrane via spectrin bound to ankyrin, which in turn has the capacity to bind to membrane proteins (*v.i.*). Microfilaments also interact with membranes through vinculin, which then binds to metavinculin, an integral membrane protein (*v.i.*). Considering the numerous membrane proteins that can re-arrange in the membrane, there are likely to be other linking proteins.

2.5 Potential microtrabecular proteins

A fourth class of cytomatrix structures cannot yet be assigned unequivocally to a single structural protein as microfilaments, microtubules and intermediate filaments can be. These structures have been called microtrabeculae; they consist of a class, heterogeneous both in length and in diameter, of projections and interconnections which frequently terminate in densities apposed to various structures [22]. From the discussion above, it is clear that many of the crosslinking proteins are of a molecular weight and diameter to be candidates for such microtrabeculae. Later discussions of particle movement within the cell will give further reasons to consider microtrabeculae as cross-linking proteins. However, it is unlikely that we have exhausted the whole range of such proteins. In addition, there may be a distinct class of proteins which crosslink the crosslinkers i.e., which bind not to the structural filaments but to the proteins bound to them, and such proteins would be additional candidates for microtrabeculae.

A striking new candidate for microtrabecular proteins is spasmin, a protein which undergoes conformational changes in the presence of calcium. Spasmin was originally described in protoza, in which spasmin provides the mechanochemical basis for contraction. As calcium is pumped out of the matrix and into reservoirs, spasmin relaxes and the cell elongates [39]. Spasmin has subsequently been detected antigenically in both the striated rootlets of flagellated cells and near the centriole of mammalian cells. If, in fact, spasmin or a related protein turns out to be widely distributed as a mechanochemical protein in the absence of other filament systems, it would provide an additional strong candidate for microtrabeculae (v.i.).

2.6 Covalent modification of cytoskeletal proteins

A large number of cytoskeletal proteins undergo covalent modification after translation. One of the commonest modifications is phosphoryla-tion. Among the proteins known to be phosphorylated to varying extents are fibronectin, filamin, myosin heavy and light chains, vinculin, β tubulin, vimentin, α actinin, desmin, α and β tropomyosin, and spectrin [40, 41]. The majority of these are phosphorylated on serine, and to a lesser extent on threonine. The only cytoskeleton proteins showing phosphorylation on tyrosine are vinculin, filamin and vimentin, although these last three are only conspicuous in transformed cells, and even at that time form between 2 and 20% of the phosphate residues [41]. In some cases, the role of the phosphate groups is known. Non-muscle myosin light chains are activated by phosphorylation [42]. In two protozoans, myosin heavy chains disassemble with phosphorylation [43]. Many microtubule-associated proteins are more extensively assoc-iated with microtubules when phosphorylated [19]. However, for the majority of these proteins, neither the functional significance of phos-phorylation nor the effect on cytoskeletal association is known.

In addition to phosphorylation, a number of other covalent modifica-tions have been detected. Tubulin is post-translationally tyrosinylated and it may be glycosylated. Actin has two covalent modifications. It contains N-methyl histidine, and an N-terminal acetyl group; the methyl-histidine is shared with myosin. Calmodulin, myosin, and *Acanthamoeba* actin contain a trimethyllysine group. Vinculin contains to varying extents an acyl group that may be a 16 carbon side chain. The functional significance of these modifications is largely unknown. Furthermore, it is quite likely that many more cytoskeletal proteins undergo covalent modifications and that a systematic search for these might reveal a pattern suggestive of the functional significance.

Our present understanding of the protein chemistry of the cytoskeleton indicates where further knowledge is needed. Clearly the stoichiometry of these proteins *in vivo* needs to be accurately measured, since it is known that some of these proteins have significant effects at stoichiometries of 1 to 500 [8]. Secondly, the association constants for these proteins need to be known in order to predict the relative occupancies when several proteins bind at the same site. Thirdly, measurement of these properties under physiological conditions (appropriate pH, pCa, NTP, ionic strength, etc.) will be important. For instance, F-actin formed under magnesium-rich conditions behaves differently both by itself and in association with other proteins, than does potassium polymerized F-actin [5, 44]. This may have physiological relevance, since in the amoeba the magnesium concentration is higher and the potassium concentration lower than in most vertebrate cells. Possibly the two forms of F-actin are relevant to different cell types. In addition, continued attention must be paid to proteolysis. In some cases, proteolysis is artifactual and can generate fragments of proteins which retain binding properties that interfere with the behavior of a native protein [6]. In other cases, as in the intermediate filament protease, proteolysis may in fact be a cellular mechanism for regulating the state of the cytoskeleton. A complete catalog of cytoskeletal proteins will probably only be generated by continued attention to all potential binding sites within a filament, the distinct ends and the sides, and by an appreciation for potential crosslinks between filament systems. Finally, all such interactions of filaments can occur either in a constitutive or regulated manner, and only critical attention to the physiological constraints on each will reveal the significance of such interactions.

3 Cytoskeletal architecture

Just as the nearly uniform genetic code allows a multitude of related species to evolve, so also does the common fundamental design of the cytoskeleton find expression in forms specific to species and tissues. The core structural proteins are in general highly conserved, but specific modifications of the accessory proteins have permitted fine tuning of the cytoskeleton for the functions of each cell. Thus most cells are characterized by a particular spectrum of cytoskeletal proteins present in a particular stoichiometry and arrangement.

The cytoskeleton confers upon the cell its shape, its ability to attach to substrates or to other cells, freedom of movement, the capacity to move material within the cell, and the capacity to export material from the cell. Such a long list of functions would present a formidable task in analyzing their structural basis individually. Fortunately, by the appropriate choice of differentiated cells, it is possible to isolate some of these functions for study.

Finally, in addition to cytoskeletal specificity as a function of cell type, the cytoskeleton can undergo regulated changes in architecture. Some of these changes are in response to particular stimuli and may either be autonomous or require either protein synthesis or RNA synthesis. Other such changes occur over the lifetime of the cell, and still other architectural changes are visible during the development of a cell lineage.

3.1 Red blood cells

The mammalian red blood cell has the simplest cytoskeleton known. It consists of a submembranous meshwork of actin and spectrin once described as 'an anastomosing framework like a net woven by a myopic fisherman.' [45]. It is almost misleading to think of actin as the structural protein, since there is one spectrin molecule for about every five actin molecules. The actin in the red blood cell is predominantly in the F-form since it contains ADP rather than ATP. The actin monomer concentration within the red blood cell is not affected by cytochalasins. The actin oligomers, which contain 10–17 molecules, display sites that behave like the fast assembly end of a filament, but no slow assembly ends are measurable [46].

The complex meshwork underneath the red cell membrane is made possible by multiple binding sites on spectrin [47]. The head-to-tail tetramers of spectrin can crosslink the oligomers of actin by binding to the sides of the actin protofilaments. In addition, the spectrin molecules contain, on the β spectrin subunit, a binding site for ankyrin. Ankyrin binds both to spectrin and to an integral membrane protein, Band 3, to form a cross-bridge between the actin—spectrin meshwork and the membrane. The actin—spectrin associations are stabilized by yet another protein, Band 4.1, which binds near the end of the spectrin molecule occupied by the side of the actin filaments. These molecular relationships are shown in Fig. 3.1.

The role of these interactions among the cytoskeletal proteins and membranes is shown by several experiments. Partial removal of spectrin from the red blood cell ghost increases the lateral mobility of the integral erythrocyte membrane proteins. This indicates that substantial constraints are placed on the membrane proteins by their association with the skeletal network. Secondly, anticytoskeletal drugs such as cytochalasins can alter to some extent the deformability of the red blood cell [48]. Finally, any of several conditions that leads to red blood cell membranes depleted of the cytoskeleton, especially spectrin, also leads to a more homogeneous distribution of membrane lipids; it has been suggested that the cytoskeleton plays a role in maintaining the asymmetric distribution of lipids in the bilayer. Defects in spectrin may be a cause of the disease hereditary spherocytosis.

Simple as the mammalian red blood cell is, it highlights the need to keep in mind the multiple interactions of most cytoskeletal proteins. Spectrin can function as it does because it has binding sites for at least four proteins on each monomer. The spectrin—actin network is substantially modified by the presence of Band 4.1 and, finally, the availability of Band 3 may be modified by the binding to Band 3 of various glycolytic enzymes [49]. Thus, for most of the proteins in the cytoskeleton, it is necessary to keep in mind the relationships to not one but several other proteins.

Mammalian red blood cells maintain their shape with an actin-based cytoskeleton. In doing so, they have dispensed with the other two

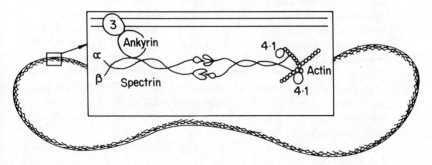

Fig. 3.1

25

filament systems; these are still present in avian and other red blood cells [50, 51]. Except in mammals, red blood cells contain a peripheral band of microtubules called the marginal band. This bundle of microtubules appears in the electron microscope to be crosslinked and is organized around a pair of centrioles. Treatment with cold will disassemble the microtubules of the marginal band. When the red blood cells are re-warmed, the microtubules begin to reform from the centriole; no other nuclei are seen elsewhere in the cell. The microtubules continue to elongate until the full circumference of the cell has been refilled. The centriole nucleates the microtubules from the pericentriolar material that surrounds the centriole triplets.

The intermediate filament network in non-mammalian red blood cells consists of vimentin filaments, polymers of a 52 kD polypeptide found in many cells of mesenchymal origin. The vimentin filaments are decorated with a 230 kD protein, synemin, which is distributed period-ically along the vimentin filaments. The intermediate filament network does not overlap with the marginal band of microtubules. However, it associates with the membrane and spans between the membranes to form a cage around the nucleus within the central region of the red blood cell. Synemin appears to crosslink this intermediate filament network [52].

Non-mammalian red blood cells also contain a submembranous actin—spectrin network and their spectrins, like those of mammals, are capable of being phosphorylated and of binding calmodulin. The func-tional significance of binding calmodulin and the phosphorylation are not known for either group of red blood cells. It is possible, however, that the calmodulin-binding and phosphorylation of the mammalian red blood cell spectrin are vestiges of the more complicated cytoskele-tal relationships of the nucleated red blood cells. Nor is it known yet what structural specialization allows a mammalian red blood cell to dispense with the two filament systems seen in all other red blood cells.

Although the red blood cell cytoskeleton appears to serve the sole function of maintaining cell shape, there are nevertheless developmental changes seen in its assembly [53]. Two of the cytoskeletal proteins, Band 4.1 and 4.2, are delayed in their synthesis until the last stages of reticulocyte maturation. Also, the spacing of synemin along the vimentin filaments changes with development from the embryo to the adult, rising from an average periodicity of 230 nm in the ten-day-old embryo to 180 nm in adult erythrocytes. Finally, the association of α and β spectrin changes with the developmental stage of the embryo; its pro-gression suggests that spectrin assembly is limited by the availability of binding sites on the membrane.

3.2 Platelets

Platelets are anucleate cellular fragments which circulate in the blood and participate in clotting. In addition to the shape maintenance that red blood cells exhibit, platelets perform the additional cytoskeletal functions of changing shape and attaching to surfaces. Platelets have a restricted repertoire of shapes. The resting platelet is a discoid, symmetrical cellular fragment. Activation causes numerous filopodia to extend. If a suitable surface is available, the activated platelet will then spread on that surface. The transition from the resting to the filopodial stage is to some extent reversible, but once spreading is initiated, it is difficult to prevent spreading to completion [54].

This two-step sequence of shape change involves two filament systems [55]. The discoid platelets contain a marginal band of microtubules crosslinked by microtubule-associated proteins. In addition, platelets contain actin and numerous actin-associating proteins; these proteins undergo co-ordinated rearrangements during filopodial extension and spreading. The three stages are shown in Fig. 3.2.

The cytoskeletal rearrangements have been revealed in detail by immunofluorescence [56]. The behaviour of the tubulin is relatively straightforward. In the resting cell, microtubules are found solely in the marginal band. During the subsequent stages of activation, they undergo distortion and rearrangement, but this appears to be a passive consequence of the activities of the microfilament system. In contrast, the actin and actin-associated proteins undergo highly specific patterns of rearrangement. In the resting platelet, actin, myosin, filamin, and α actinin are visible in a granular pattern that suggests large, possibly loose, aggregates, but the tropomyosin is very nearly uniformly distributed. During filopodial extension, actin is found concentrated along the length of the filopodia and somewhat excluded from the central domain. Tropomyosin and filamin are also found along the entire length of the filopodia, but tropomyosin is altogether excluded from the central region, while filamin is present in a loosely filamentous pattern. In contrast, myosin and α actinin are only found in the proximal half of the filopodia, and are present in the central domain in a

Discoid

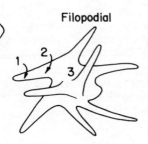
Filopodial

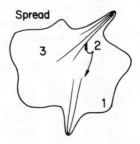
Spread

Fig. 3.2

granular or loosely fibrillar pattern. Spreading leads to further rearrangements. Actin is now at the borders of the spread platelet and occasionally concentrated in bundles that penetrate toward the central region; it is less conspicuous in the central domain. Myosin is found predominantly in the central region, as is tropomyosin, although the patterns of the two are not identical. Filamin also appears to be distributed filamentously in the central domain. In contrast, α actinin is largely excluded from the central region, and is highly concentrated at the periphery. These rearrangements are summarized in Table 3.1. The distribution of other actin-associating proteins, such as profilin and the 90 kD protein, is not known.

This complex and co-ordinated rearrangement is the setting in which the known biochemistry of platelet proteins must be explained. Calcium potentiates the phosphorylation of myosin light chain [42] and potentiates profilin binding to actin [5]. It inhibits the filament crosslinking of non-muscle α actinin [5]. In contrast, both filamin and tropomyosin are calcium insensitive in their binding to F-actin [5]. At first, this pattern of calcium sensitivity is counter-intuitive since platelet activation is accompanied by calcium influx and a rise in pH. However, some of the rearrangements observed might be explained by the following sequence of events. In the resting platelet, actin is sequestered in an association with a calcium-sensitive actin-binding protein. When calcium flows in and the pH rises, profilin dissociates the actin from this protein and frees it for polymerization onto a calcium-insensitive nucleating protein or a protein which nucleates only in the presence of calcium. This protein would presumably be less numerous than the sequestering protein. As the filaments elongate to more than six monomers, tropomyosin could bind to them and filamin could begin to crosslink the filaments. The myosin light chain being phosphorylated, it would now be able to interact with the actin filaments.

Several steps of this proposed sequence are known to occur. There is a calcium influx and change in pH [54]. Profilin binds more strongly to actin in the presence of calcium and can polymerize onto the fast

Table 3.1 Platelet cytoskeletons

Region	Discoid 1	2	Filopodial 1	2	3	Spread 1	2	3
Tubulin	++	−	−	−	+	±	−	tangles
Actin	granular	++	++	±		+	++	±
Myosin	granular	−	+	granular		±		+
α Actinin	granular	−	+	±		++		±
Filamin	granular	+	+	±		±		+
Tropomyson	diffuse	++	++	−		±		+

assembly end of actin filaments [5]. Non muscle tropomyosin binds to six actin subunits in the filament, and filamin does not appear to interact with G-actin. It is also known that in the resting platelet actin is found in aggregates 10–20 nm in diameter and 20–40 nm long; too large to be actin–profilin complexes. These aggregates release α actinin upon incubation in calcium.

To substantiate this pathway further, the affinity coefficients for the various proteins and complexes would need to be measured at the physiologically relevant calcium concentration and pH, keeping in mind that F-actin polymerized with potassium differs biochemically from that polymerized by magnesium. It should be noted, however, that even this pathway does not account for the fact that no two proteins within the platelet have the same distribution in the filopodial and spread forms. It is possible that still unexplored factors, such as the interaction with phospholipid bilayers, need to be explored.

The platelet, then, represents a two-filament system, in which the two filament systems undergo minimal interaction except for the passive distortion of the microtubule ring by the actin networks. It is clear, however, that the platelet provides an excellent model for the recruitment of a family of actin-associating proteins into a specific and dynamic microfilament structure.

3.3 Fibroblasts
Fibroblasts secrete extracellular matrix and contribute to tissue compaction. They play a role in wound healing and, during embryogenesis, fibroblast-like cells undergo numerous movements within the embryo to give rise to several mesenchymal derivatives. Thus to the functions of maintaining cell shape, undergoing a single stereotypical change in shape, and spreading on a substrate, fibroblasts add the additional cytoskeletal functions of active cell locomotion, polarization and generation of tension. In addition, since they are nucleated cells, they locate material within the cell and direct other substances for export. This larger repertoire of cytoskeletal functions is reflected in the increasingly complex cytoskeletal organization.

The organization of the fibroblast cytoskeleton is largely a function of cell cycle stage and available substrate. Thus the rearrangements seen after replating cultured cells are comparable to those seen after mitosis, or to those that occur during embryogenesis or wound healing. However, the cultured cells are substantially more accessible, both to observation and to experimental intervention.

A rounded fibroblast responds to contact with a suitable substrate by the extension of numerous filopodia. These long, slender filopodia appear to probe the immediate environment of the fibroblast (*v.i.*). If they touch a substrate, an attachment may be formed. If they make contact with a non-adherent particle, frequently the filopodium will attach to the particle and retract with it. As soon as a sufficient number of contacts have been made, the cell begins to ruffle around its periphery; the ruffling and filopodial extensions may interconvert into each other. At this stage, the actin is found extensively associated with the filopodial ruffles and in a number of transnuclear cables [57]. As the cell continues to spread, these transnuclear cables reorganize and begin to be distributed in the interior parts of the cell as a network of intersecting polygons. Over a period of hours these polygonal networks continue to be reorganized into stress fibers, until the characteristic interphase appearance of the cell is attained.

Tropomyosin follows a somewhat different pattern of rearrangement. At the early stages, when actin is extensively associated with ruffles and in a cluster of transnuclear cables, tropomyosin is almost exclusively in a diffuse non-filamentous pattern around the nucleus. When polygonal networks have been established, tropomyosin is found associated with them but excluded from the polygon vertices. After rearrangement, tropomyosin is nearly continuous along the length of stress fibers, but it displays faint periodicities at an approximate spacing of $1.5\ \mu$.

A third pattern of rearrangement is seen for α actinin. At the earliest stages it shows, like tropomyosin, a central diffuse location. By about eight hours, the α actinin is found in patches that coincide with the vertices of the actin polygons. These patches are focal contacts, regions in which the cell approaches the substrate to within less than 15 nm. In the fully reorganized cell, the α actinin is found along the length of stress fibers, again with the periodicity of about $1.5\ \mu$, but antiperiodic to tropomyosin. At this stage, α actinin is found both along the length of stress fibers, and is also highly concentrated in the ruffling membranes.

Several other actin-associating proteins are located in the fibroblasts [50]. Myosin is found predominantly along stress fibers in a pattern that coincides more or less with that of tropomyosin. Myosin is absent, however, from microspikes (small slender projections from the cell), ruffles and focal contacts. One of the few proteins that is co-extensive with actin is filamin. The only actin-containing structure lacking filamin is the extreme tip of the microspike. Filamin is found in spaces between stress fibers as well, opening the possibility that it associates with other proteins within the cell.

Two actin-associated proteins with particularly striking distributions in the fully spread fibroblast are fimbrin and vinculin. Fimbrin (68 kD) was originally isolated from microvilli (*v.i.*). Although small amounts of fimbrin can be detected along stress fibers, fimbrin is concentrated at the periphery and is abundant in ruffles, microspikes, microvilli and filopodia [59]. Vinculin, by contrast, is predominantly associated with focal contacts, with a small remainder in a central diffuse pattern. Vinculin will remain associated with the cytoplasmic face of the cell membrane in a focal contact even after the actin of the focal contact has been removed by various treatments [60]. It is therefore believed that vinculin is one of the proteins closest to the plasma membrane in the focal contact.

It is clear that actin participates in many structures within the fibroblast cytoskeleton; each of these structures is characterized by a particular spectrum of associated proteins. A few of these structures appear to be homogeneous or isotropic, although it has been suggested that stress fibers are a consequence of ordering a homogeneous network by tension. Similarly, filopodia resemble, in some respects, actin bundles. Nonetheless, any close examination of the fibroblast cytoskeleton presents the pressing question of why different actin-associated proteins are restricted to different domains within the cell. Some, such as vinculin, may be so restricted because of additional binding capacities, for example, to the membrane. Whether this will present an adequate explanation in the long run or whether additional dynamic interactions are required can only be determined by further research.

The second major filament system in fibroblasts is that of the microtubules. These converge upon a central region in the cell focused around the centrioles. Immediately after plating, no complex network of microtubules is obvious. However, with time the microtubules lengthen, develop bends and sinuousities, and eventually reach from the centrosome to the cell periphery [61]. Microtubules are also found during mitosis and sometimes in the primary cilium, a vestigial flagellum-like structure. Both of these will be discussed in Chapter 4. During interphase, microtubules appear to play a role in cell polarization, i.e. the ability to generate ruffles and filopodia on one side of the cell and to undergo net locomotion. They are also involved in moving materials from the Golgi to the exterior for the extracellular matrix. Both of these behaviors will be discussed later.

The third major filament system in fibroblasts is that of the vimentin intermediate filaments. Intermediate filaments are found weaving throughout the central region of the cell and extending towards the periphery. They extend after mitosis only after microtubules elongate. In addition, vimentin fibers surround the nucleus and make close contact with stress fibers [62]. Although the intermediate filaments of fibroblasts are generally composed of vimentin, there is at least one well-documented case of fibroblasts from heart containing small amounts of desmin, usually found in muscle cells. In these cardiac fibroblasts, the desmin appears to be co-polymerized with the vimentin intermediate filaments [63].

For all these proteins, our primary evidence for their location is immunocytochemistry. Reliable answers using this technique depend upon both the specificity of the antibody and the accessibility of the protein within the cytoskeleton. That the evidence from immunofluorescence studies is generally reliable is largely confirmed by micro-

31

injection studies, in which a cell is injected with fluorescently labeled proteins. Such studies have been performed for α actinin, vinculin, tubulin, MAPs, and actin (*v.i.*). So far no structure has been illuminated by microinjected proteins that was not already shown by immuno-fluorescence to contain that protein, thus confirming the specificity of immunofluorescence. However, there may yet be a subpopulation of structures too stable or dense to be accessible to proteins in solution or to antibodies.

The fibroblast cytoskeleton can be studied at high resolution in the electron microscope. Some immunocytochemistry has been conducted at the EM level. In addition to these specific proteins, further detail can be seen either by use of extracted cytoskeletons or by appropriately fixed whole cells. When fibroblasts are extracted in the absence of an osmoticum, many structural fibers persist and can be identified by immunoferritin techniques [64]. Actin filaments associate with other actin filaments, or with microtubules or intermediate filaments. In addition to the three characterized structural filaments, these somewhat stringent cytoskeletons also reveal numerous heterogeneous filaments that crosslink the major filament systems. Under gentler extraction conditions, in which sucrose protects the cell during extraction, a yet more complex nextwork can be identified ([65], Fig. 3.3). In such a

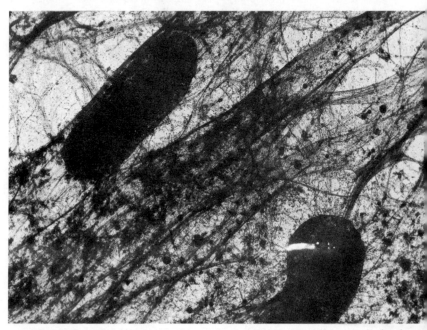

Fig. 3.3 Transmission electron micrograph of fibroblast cytoskeletal framework.

cytoskeletal framework, filament frequency is so great and in some cases the diameter is sufficiently small, that conventional thin sections fail to reveal them. Finally, when intact cells are examined, either in thick section or as cells grown on EM grids penetrated by high voltage electrons, further complexity is revealed, including the tenuous, variable microtrabeculae which associate with both the known major filament systems and various intracellular organelles. The increasing complexity of filamentous structures as a function of the protection given the cytoskeleton during preparation may reflect the characteristic residence time for proteins on the cytoskeleton. Thus those proteins which have brief, if frequent, associations can only be preserved by techniques which stabilize that association, whereas techniques which allow extensive extraction will predominantly reveal those proteins which rarely exchange with the soluble phase of the cell.

3.4 Muscle

Muscles are mesenchymal cells that have specialized for contraction. The architectural specialization of the cell depends upon the mode and strength of contraction it will perform. Thus muscles present a series of cells, from some smooth muscles which are only slightly more specialized than a fibroblast, through to the archetypal muscle, skeletal striated muscle, one of the most geometrically organized cells in the body. This range of functions is very old, since both smooth and striated muscle can be found in all the animal phyla through to the *Coelenterata.*

Smooth muscles, closest in organization to fibroblasts, are long, slender, spindle-shaped cells with a single, central nucleus. The cell membrane displays numerous dense bodies that lie immediately underneath the membrane and attach the smooth muscle either to extracellular matrix or to other smooth muscles. These dense bodies are rich in vinculin, and numerous microfilaments and some intermediate filaments anchor into them. Microfilament bundles crisscross the cell obliquely so that in the relaxed state they are birefringent; during contraction they form a lattice work which is less highly ordered and no longer birefringent (Fig. 3.4). Myosin thick filaments are not conspicuous in smooth muscle cells. Thick filament assembly may be regulated by the phosphorylation of myosin light chains, which is mediated by a myosin light chain kinase [42]. Microtubules are present in the adult smooth muscle cell, but they are more conspicuous in smooth muscle cells in culture, in which they have the organization seen in fibroblasts, i.e. long sinuous fibers radiating from a central location [66]. The biochemical differentiation of smooth muscle cells may account for the differences in architecture between fibroblasts and smooth muscle cells. Smooth muscles contain α and γ actin isoforms characteristic of smooth muscle, as well as the β and γ isoforms of actin found in fibroblasts. There are myosin heavy chain and light chain isoforms specific to smooth muscle as well [67]. A strikingly interesting biochemical

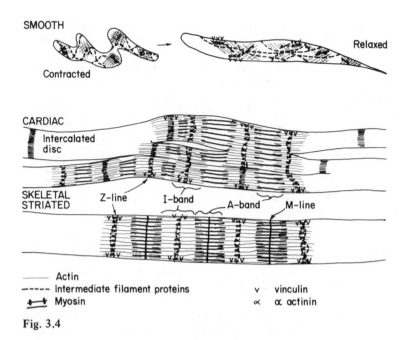

SMOOTH

Contracted

Relaxed

CARDIAC

Intercalated disc

SKELETAL STRIATED   Z-line   I-band   A-band   M-line

——— Actin
----- Intermediate filament proteins   v   vinculin
+——+ Myosin   ∝   α actinin

Fig. 3.4

heterogeneity in smooth muscle cells, however, is that of the intermediate filament proteins. Smooth muscle cells may contain desmin filaments, as do sarcomeric muscles, or vimentin, or both filaments at the same time. This dual expression is seen both in primary culture and *in situ* in the adult animal; the heterogeneity in the adult shows a graded expression that suggests that which intermediate filament is expressed may be partly determined by whether or not the smooth muscle is extensively engaged in contraction [68].

Considerably more is known about the cytoskeletal architecture of the sarcomeric muscles, cardiac and skeletal striated muscle. These muscles resemble smooth muscle in having a large fraction of their interior occupied by microfilaments and accessory structures, and in devoting a significant fraction of cellular protein to the contractile proteins. However, both heart and skeletal muscle are more highly organized, so that thin filaments are in register, with intervening bands of myosin thick filaments. The thin filaments terminate on an electron dense line called the Z line; myosin thick filaments center on the M line. Each subunit from Z line to Z line is called a sarcomere.

Cardiac muscle cells, like smooth muscle cells, are mononucleated, but their microfilaments are more highly organized without the extreme

34

regularity seen in skeletal muscle ([69] Fig. 3.4). The increase in order is visible from the surface inward of the cardiac cell membrane. A cortical lattice contains β spectrin and vinculin, organized into rows which flank or overlie the Z line [70, 71]. These rows of organized proteins, called costameres, are spatially related to the sarcomeres and form ribs around the myofibril. Costamere spacing varies with the spacing of the sarcomeres during contraction. Another organized surface structure in cardiac cells is the intercalated disc, found at the domain in which myofibrils are maintained in register from one cell to the next, which in effect replaces a Z line with two cell membranes. At the intercalated disc, α actinin, actin and vinculin are found and around them desmin is highly concentrated [72].

The interior of the cardiac muscle is largely occupied by myofibrils that branch; thus each muscle cell forms contacts via intercalated discs with several other cardiac cells. The myofibrils consist of a series of sarcomeres. The Z bands contain α actinin and actin in a central domain surrounded by desmin filaments. The thin filaments that radiate away from the Z discs contain the cardiac α actin filaments and decorated with the cardiac forms of troponin and tropomyosin. These thin filaments are not of identical length, and vary by as much as 0.6 μm [69]. Thin filaments interdigitate with myosin thick filaments that center on the M line. Thus a large fraction of the cardiac cell interior consists of an anastomosing network of actin and myosin filaments held in a spatial array that permits evenly distributed tension to be transmitted to neighboring cells. The conspicuous presence of desmin filaments in cardiac cells is consistent with the role suggested in those smooth muscle cells that contain large amounts of desmin, i.e. that desmin is involved in anchoring together muscle cells in some way against the tension that they themselves generate.

Striated muscle shows the greatest degree of spatial organization of the three muscle types (Fig. 3.4). This organization is also visible from the outside in. Like the cardiac cells, striated muscle cells display costameres on their inner surfaces. The striated muscle costameres are known to contain γ actin, spectrin, intermediate filament proteins as well as vinculin, again organized as in the cardiac muscle, in stripes that parallel the Z line and expand and contract with the myofibrils [73, 74]. Going in from the surface, desmin filaments form strands peripheral to the Z bands and appear to connect the Z bands to the adjacent sarcolemma. The desmin filaments surround but do not penetrate the central domains of the Z band discs. Present in the central domain, as in cardiac cells, are α actinin, actin, and additionally Z protein. The periphery of the Z disc contains, in additon to desmin, the intermediate filament-associated protein synemin, the actin-binding protein filamin, and small amounts of spectrin [75, 76]. Radiating out at 90° from the

Z line are the thin filaments of the I band, composed of striated α actin, the skeletal muscle forms of troponin T, C, and I, and tropomyosin. In a restricted domain of the I band is the N line which contains at least the protein nebulin, a very large myofibrillar protein, approximately 500 kD [77].

Interpenetrating the I band thin filaments are the myosin thick filaments which contain the myosin heavy chain and light chain, and C protein of 140 kD. C protein exhibits calcium-sensitive binding to the I band in addition to binding to myosin, and exists in several isoforms. Molecularly distinct species have been identified that are specific to the pectoralis major (fast twitch) and the anterior latissimus dorsi (slow tonic) muscles [78]. Thus, as for the myosin heavy chains, isoforms are specific to the physiologically different fibers. In addition, however, some fibers show heterogeneity in their C protein, even within a single sarcomere. Thus the PLD muscle contains C protein of both forms, present in each sarcomere. At the center of the thick filaments is the M line which contains 165 kD myomesin. In additon, the muscle specific form of creatine kinase, MM–CK, is also present at the M line [79].

The architecture of the myofibril and presumably of the costameres is dedicated to the contractile function of the muscle cell; this has focused attention away from the non-contractile aspects of striated muscle. However, all of the other cellular organelles are also localized, although not with the near crystalline regularity of the sarcomeres. One of the more intriguing aspects of striated muscle architecture is that a γ-specific antibody recognizes γ actin as decorating mitochondria but not participating in the I bands of sarcomeres. Gamma actin is also present in a subcortical layer underneath the plasma membrane [80].

It is important to note, however, that any individual striated myofibril rarely exhibits the crystalline appearance of a well-drawn diagram. Rather, muscle is a statistically regular structure. For example, filament length varies within a single sarcomere for thin filaments, so that within one sarcomere there may be variations of $0.18-1.2$ μm [69]. Additionally, neither the Z lines nor the M lines in adult muscle are drawn with the regularity of a ruler. Thus models for muscle assembly cannot be direct extrapolations of bacteriophage assembly, i.e. they cannot be a strictly linear and deterministic pathway with a single, definable endpoint. The end-product of assembly in muscles is a highly, but not perfectly, regular structure; models of sarcomere assembly must reflect this fact.

The great regularity of muscle has attracted attention to muscle development as a model for the generation of spatial order. The rearrangements during development of several muscle proteins are known in detail. Alpha actinin is first found as in fibroblasts or smooth muscle cells, i.e. in punctate patterns along stress fibers. Around day 4, as sarcomeres begin to organize, it is found in the central domains of the

Z discs. Very early in muscle culture, filamin is also found distributed as in fibroblasts, i.e. along stress fibers. Filamen then disappears for several days; it reappears at the periphery of Z discs. During this hiatus, there has been a switch in expression; that is, the filamin expressed earlier in culture is biochemically similar or identical to that found in smooth muscles and fibroblasts, but the later appearing filamin, found associated at the periphery of Z discs, is a distinct biochemical species [81]. Desmin is not found until the muscle cells cease dividing and begin expressing the muscle-specific program. Only a small amount of desmin is found in mononucleated myoblasts. It is principally expressed in the multinucleated myotube. For the first several days of expression desmin forms a diffusely filamentous pattern that resembles that of vimentin, for instance, in a fibroblast. By about day 8, a day later than filamin and several days later than α actinin, desmin begins to associate with the periphery of the Z discs and is visible in register with the sarcomeres.

Muscles at all stages of development contain myosin. However, in the earliest stages, myosin is of the cytoplasmic variety found in fibroblasts. This myosin is distributed as in fibroblasts: in stress fibers with a faint periodicity, and in a submembranous rim around the cell. This pattern continues in the developing myotube, although stress fibers are predominantly at the ends of cells. The muscle-specific myosin occurs in the myotube center either in fibers without periodicity or in sarcomeres, depending on the maturity of the muscle [82, 79]. Myosin heavy chain changes its organization over time within the muscle, since monoclonal antibodies that demonstrably recognize the same myosin heavy chain none the less show a developmental progression of patterns within embryos of different ages [83].

Two non-contractile proteins are important for the correct development of muscle. Tubulin and microtubules are present at all times during muscle development. These microtubules are generally oriented along the long axis of the developing muscle cell. Perturbations of microtubule function perturb muscle development, and will be discussed later. A second non-contractile protein is vimentin, present in the young muscle cell and whose presences continues in a diffuse pattern even after sarcomeres containing α actinin have begun to organize. The organization of vimentin in muscle is a matter of some dispute. One investigator found vimentin organized around the Z discs, around the time that desmin occurs also [84]. However, two other laboratories did not detect vimentin organized in the sarcomeric pattern [85, 86].

It is worth examining the possible sources of this disagreement as a measure of the strengths and weaknesses of immunofluorescence for detecting a protein. An immunofluorescence pattern indicates the presence of one or several proteins: (1) the antigen of interest, (2) a related but biochemically distinct antigen, (3) an antigen that contaminated the original preparation used to raise the antibody, or (4) a highly adsorbent site that is not specific for any antibody. A negative result, i.e. the absence of a pattern, can result from several conditions: (1) absence of antigen, (2) inaccessibility of antigen, or (3) overly stringent staining conditions. There is no single procedure that absolutely eliminates false positives or false negatives because the controls needed to increase the reliability of either a positive or negative conclusion are themselves susceptible to the same constraints upon the original staining reaction. Thus, the conflict about vimentin in the myofiber could result from any of the following. (1) A muscle-specific form of vimentin, present at all Z discs in culture and in adults, is detected by one group's antibody because it is a broadly-specific antibody, but not by the other groups' because theirs recognize only fibroblast vimentin. (2) No vimentin is present at Z discs at any time, but a pattern appears there because the vimentin antibody used has a low affinity for desmin, or has a high non-specific affinity for Z discs which has not been reduced by the appropriate choice of staining conditions. (3) There is a genuine developmental difference between the two sets of muscles examined, i.e. the vimentin-containing Z discs result from systematic depletion of fibroblasts, and an immature state or an artifactual state has been produced by the increased purity of the muscle cells. Which of these three possibilities is in fact responsible can only be resolved by an exchange of antibodies and application under both conditions.

Muscle cells have specialized their cytoskeleton for contraction and have extensively proliferated the microfilamentous component to fill the cell interior, not only the cell cortex. To do so, muscle-specific isoforms are used in forms and patterns specific to each of the major muscle classes. To what extent the differences in architecture are the consequence of biochemical differences in the component proteins is not yet known. However, it is clear that for all muscle classes, including skeletal striated muscle, the cytoskeleton is a dynamic structure which develops over time and continues to be regenerated and modulated in response to use.

3.5 Epithelial and endothelial cells

Epithelial cells perform many roles in the body. They cover the skin with a strong and waterproof layer, line the gut with a resilient and adsorptive layer, and form diverse glands within the body. In light of these many functions it would be surprising if the epithelial cytoskeleton were uniform. However, just as muscles display a progressive specialization and expansion of one cytoskeletal system, the stress fiber, so also epithelial cells can be largely characterized by their elaboration of one or both of two particular cytoskeletal systems: the intermediate filament networks and the cortical microfilament structures.

Epithelial cells resemble fibroblasts in several respects. As epithelial cells spread on a sub-stratum, they form focal contacts between the cell and the surface. These focal contacts, rich in α actinin, correspond to the ends of microfilament bundles, as in fibroblasts. However, the pattern of the focal contacts and bundles differs; the bundles are generally shorter, and the focal contacts, largely restricted to the cell edge, are less readily resolved than in the fibroblast [87].

The detailed structure of stress fibers in epithelia has been studied by immunofluorescence. The results show that the epithelial cell–cell junctions, attachment plaques, and stress fiber foci all contain α actinin, as in fibroblasts. However, in the epithelial cells, the tropomyosin and α actinin are more closely spaced than in fibroblasts, as are the periodic densities that underlie the stress fibers [88]. Myosin is found periodically among the stress fibers [89]. The myosin distribution is again a function of cellular activity. During spreading, actin was found alone in the surface blebs and ruffles but with myosin at the bases of these structures. The more spread and less motile the epithelial cells, the more extensively the actin and myosin are found restricted to the stress fibers [90].

Epithelial cells contain extensive microtubular networks which, as in fibroblasts, frequently connect with subcellular organelles. Two to four nm filaments extend from the microtubules and often terminate on pigment granules or other structures. Substantial cell disruption can occur before separating such granules from the microtubular network [91]. In one epithelial system, the polarity has been determined for all of the cellular microtubules, i.e. those present during interphase. The rapidly growing end of the preferred end for assembly was found distal to the cell center [92]. The implications of this for bidirectional transport will be discussed later.

The third major filament system, the intermediate filament network, is most diverse and most highly specialized in epithelia. Epithelia characteristically contain cytokeratins, intermediate filament proteins that can be divided by molecular weight and tissue distribution into at least seven major classes. These different polypeptides are products of distinct genes, not differentially processed polypeptides [86]. So far, no epithelial cell has been found that contains a single cytokeratin. All contain at least two different ones and many contain more than that. A cell's keratin composition is a function of the cell type, its growth environment, its stage in histological differentiation and the period of embryonic or later development [86]. Cytokeratins are the first intermediate filament detected, since they are found in the trophectoderm of the embryo [93]. Several features of epithelial differentiation *in vivo* are characterized by specific keratins. Thus a 50 kD and 58 kD keratin are found only in stratified epithelia, while a 56.5 and a 65–67 kD group of keratins are restricted to keratinized epidermis. In contrast, one or another of a 40, 46, and 52 kD group are found in all epithelia [94]. This specialization of keratins occurs even in closely situated subpopulations of a single tissue. Thus, the epithelium of the outer root sheaths of the hair follicle shares three major polypeptides with the epidermis between hair follicles, but lacks three others and has a characteristic 46 kD cytokeratin of its own. A 48 kD cytokeratin of the hair follicle is also found in the sole. Although close to the hair follicle and the interfollicular skin, the eccrine sweat gland epithelium has three unique cytokeratins. Thus the skin consists of many microdomains characterized by both different morphologies and differentiation of their cytoskeleton [95].

Cytokeratins can vary extensively with the growth conditions in culture. Clonal cell lines from mammary gland epithelium grow with or without hormones. When grown with the appropriate hormones, the cells grow in monolayers of loosely-packed cells, the intercellular spaces spanned by desmosomal bridges. The cytokeratin fibers bundle and anastomose in extended arrays throughout the cytoplasm and terminate at the desmosome. The four major cytokeratins in these cultured cells are those found *in situ* including phosphorylated components. These epithelial cells grown with hormones do not produce vimentin. When grown without hormones, the cells express a different set of cytokeratins and a different morphology, but do form desmosomes; they now make vimentin [96]. This latter observation may explain why two common established epithelial cell lines, HeLa and PtK, contain both cytokeratin and vimentin fibers. In these cells, as in the cultured mammary epithelium, the vimentin and cytokeratin fibers intertwine and anastomose but do not coincide. These established cells still make desmosomes, although in reduced numbers. When the cells are treated with colchicine to disassemble microtubules, their vimentin filaments form a juxtanuclear cap, while the keratin filaments, although somewhat disrupted, remain anchored at the desmosomes and largely extended throughout the cytoplasm, further evidence that in these cells the

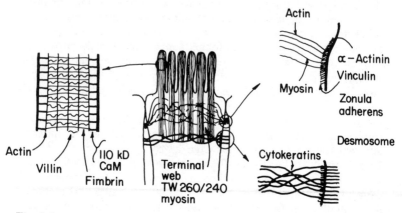

Fig. 3.5

40

two groups of intermediate filament proteins are not co-polymerized [97]. Thus, cytokeratins are diagnostic of epithelium; the particular cytokeratin polypeptides present are characteristic of the location and function of the particular epithelial cell.

The other major cytoskeletal specialization of epithelial cells, present to a greater or lesser extent, is amplification and specialization of the cortical actin meshwork and its microvilli. One of the classical sources of material for studying these structures is the intestinal epithelium which absorbs nutrients and contributes to the mechanical strength of the gut lining. The luminal face of these cells terminates in numerous microvilli anchored in a terminal web (Fig. 3.5). The microvillus core is an actin bundle of approximately 20 filaments crosslinked to each other and to the microvillar membrane. This bundle of actin filaments is capped at the distal end; proximally, it penetrates the terminal web of the cell. The structural roles in the microvillar core are filled by actin, villin (95 kD) and fimbrin (68 kD), which crosslink the actin filaments, and a 105 kD protein which binds calmodulin and appears to crosslink the bundle to the membrane [98].

These bundles of microfilaments penetrate into and are anchored in the terminal web of filaments, which consists of actin and tropomyosin, myosin in short thick filaments, and slender crosslinking filaments related to the spectrin family. Originally called TW240/260, these spectrin-like molecules share an α-subunit with spectrin but their 260 kD subunit is not particularly similar to β spectrin. The $\alpha\beta$ dimer forms thin ($\sim$ 5 nm) fibers that link actin rootlets to adjacent rootlets and the plasma membranes. An additional fibrillar component in the terminal web is a meshwork of 10 nm keratin filaments that anastomose and terminate at desmosomes along the lateral face of the cells.

This two-fold organization of the cortical actin meshwork is not unique to intestinal epithelial cells. Another group of epithelial cells which show this mixture of peninsular actin bundles and submembranous meshwork are cochlear hair cells, which extend stereocilia. The stereocilia contain a tapering bundle of actin filaments which are cross-bridged to each other and to the cell membrane. Some of these filaments penetrate into the cell and form a rootlet which is anchored into a submembranous meshwork that contains microfilaments, and fine 3–4 nm filaments that may represent a member of the spectrin family. These stereocilia are altered in their architecture by exposure to loud noises, possibly by depolymerization or fragmentation of the actin filaments at the stereocilium base where it pentrates into the subcortical meshwork and by a decrease in cross-bridges between actin filaments, reducing the rigidity of the structure [99].

The filament packing in the stereocilium has led to a striking insight into constraints on assembly for helical structures. Although these actin filaments are clearly in register when viewed longitudinally, transverse sections show a liquid packing. This irregularity in

41

transverse orientation results from the multiple equivalent binding sites around the helix. Thus, for any given cross-bridge, there are several possible binding sites for the next cross-bridge. Therefore, in all except the most highly crosslinked case, the exact pattern of cross-bridges cannot be predicted. When all possible cross-bridges have formed, the paracrystalline array of filaments results in a hexagonal packing. In the animal, a progressive change occurs; at earlier times in development a liquid order is seen, and at later times a hexagonal packing is more frequent [1]. One striking aspect of this is that the liquid order depends solely upon the helicity of the actin filaments. Therefore all helical structures, including microtubules and intermediate filaments, have a comparable freedom of structure formation; that is, having a helical substructure, they offer multiple equivalent bonding sites when forming cross-bridged structures. Thus the packing of any one such structure cannot be predicted in detail, only statistically.

A somewhat more complicated example of a peninsular structure anchored by the cortical mesh in an epithelial cell is found in the retinal epithelial cell [100]. Here, the rod inner and outer segments that project out of the cell body contain both microtubules and microfilaments. The microfilaments produce changes in length with changes in light conditions. The rod inner and outer segments contain several calmodulin-binding proteins in their cytoskeleton, including one that appears to be α spectrin. The rod is anchored into the cell body by cross connections to a meshwork of intermediate filaments, interwoven with extensive bundles of circumferential microfilaments at the level of the zonula adherens. Myosin is probably present in this meshwork, since the isolated polygons contract in the presence of ATP, and the contraction is inhibited by modified myosin [101].

Epithelia are also extensively involved in glandular structures; the cytoskeletal architecture of such structures is not as well known as yet. One secretory epithelium characterized in part is the rat liver, in which there are numerous cross-bridges between the $\beta\gamma$ actins and the cytokeratin filaments. These actins are isolated with the desmosomes in plasma membrane fractions [102].

Because epithelial cells have well organized and characteristic cytoskeletons, they have been frequent subjects for new approaches for visualizing the cytoskeleton. Intact PtK cells have been examined by high voltage microscopy after either freeze drying or freeze substitution. The overall appearance of such frozen treated cells when compared to those conventionally fixed is not markedly different [103]. In both cases numerous microtrabeculae anastomose and branch, making contact with and covering various fibrillar elements within the cytoplasm, as well as touching subcellular organelles. The complexity of such structures is largely preserved if BSC cells are extracted very briefly with Brij, although no direct biochemical data are available for the exact extent of extraction with this treatment. A somewhat less complex interior structure is seen when sucrose is included as an osmotic protectant during extraction [22]. If sucrose is left out, the resulting cytoskeleton is generally reduced to bare filaments, which in many cases show some disruption [104]. Another approach has been to fix cells with a mixture of glutaraldehyde, tannic acid and saponin. This process, which presumably involves a partial extraction of cytoplasmic proteins during fixation, acts to mordant the cytoplasmic filaments so that even in thin section, filamentous structures are more visible, an advantage which in part offsets its necessary thickening of structural detail [105]. The difficulty of seeing filaments in sections is highlighted by comparing techniques which do or do not remove embedding material (Fig. 3.6). Some attempts have been made to make use of cryosections of an intact liver epithelium to reveal cytoskeletal structure *in situ*. Although the cellular structure has suffered somewhat from crystal formation, nonetheless the desmosomal bridges and the anastomosing filaments in the cytoplasm are strongly suggestive of the cytoskeletal structure seen in cultured cells [106].

Epithelial cells show a range of specializations in two aspects of the cytoskeleton, the intermediate filament network and the actin cortical meshwork. Why these cytokeratin networks have a minimum of two subunits is not yet understood. Whether the myosin so frequently found in the meshwork actually produces contraction *in vivo* or merely maintains tension isometrically has also to be determined.

Endothelial cells, like epithelial cells, form sheets. Their functions and the physical constraints placed upon them by their situation within the body are sufficiently different that there are significant differences between endothelial and epithelial cytoskeletons. The first and striking one is that their intermediate filaments are made solely of vimentin [86]. Cultured endothelial cells frequently contain stress fibers with the normal complement of associated proteins. *In situ*, however, actin is predominantly associated with the periphery of endothelial cells as a diffuse network. Stress fibers are seen only below major branches in the arteries, sites that probably undergo the maximum mechanical stress. Stress fibers appear in endothelial cells forced to regenerate. As the cells spread and move to fill a wound, stress fibers develop extensively and are oriented toward the denuded area. During wound healing, the stress fibers follow the direction of blood flow and persist well after the naked basement membrane has been covered [107, 108].

Spreading in endothelial cells resembles that seen in fibroblasts and epithelial cells. During an initial attachment stage, filopodia extend from the cell surface and explore the immediate area; flattening follows as a greater fraction of the cell surface comes into contact with a substrate. After sufficient spreading, microfilaments begin to bundle into stress fibers and microtubules extend from the cell center in radial tracks. As soon as microtubules have extended sufficiently and formed lateral associations, various organelles move out along the radial tracks

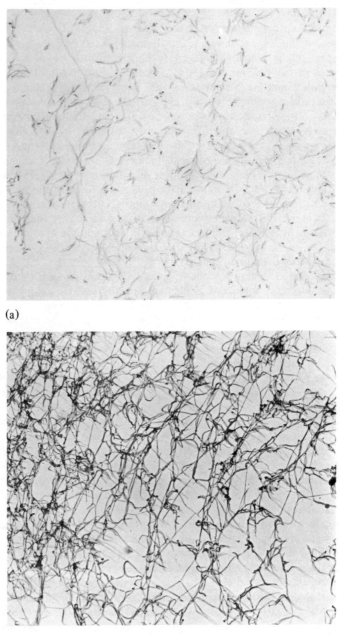

(a)

(b)

Fig. 3.6 Transmission electron micrographs of collagen gels with (a) and without (b) embedding medium present; fibrils are obscured by embedding medium.

44

of microtubules [109]. This association between organelles and micro-tubules becomes less obvious as reorganization progresses, since the microtubules become more evenly distributed. However, it is a structural correlate of the process of intracellular movement, to be discussed later.

3.6 Transformed cells

In general, transformed cells have a disorganized actin based network [110, 111]. Since a properly functioning actin cortex is necessary for efficient spreading and locomotion, it is therefore difficult to exclude the possibility that significant defects exist in the other two major fila-ment systems. However, in some cases, rather specific modulations can be detected. One spontaneously transformed cell line possesses mutant actin; a further transformant from this mutant has doubly mutant actin. The mutated actin was expressed at the same rate as the normal actin but was less highly corporated into the cytoskeleton [112]. Actin-associated proteins may be modified as well; α and β tropomyosin may be diminished, or vinculin may be more highly phosphorylated. When a temperature-sensitive virus transforms fibroblasts that express β and γ cytoplasmic actins, and some smooth muscle α actin, the α actin is specifically reduced, apparently because of a lower rate of synthesis [113].

In epithelial cells, several proteins change location with viral trans-formation. The kinase pp60^{src} is located at adhesion plaques, both in contact with a substrate and at cell—cell junctions (114). This location at adhesion plaques places it in close conjunction with vinculin, which increases in its tyrosine phosphorylation with transformation [41]. In these cells, as in normal cells, α actinin and actin are also found at the adhesion plaques, but in smaller amounts. In addition, the number and size of the adhesion plaques is less than that of the normal cell. Since pp60^{src} is a viral form of a normal cellular protein, it seems likely that this architecture is a modification of one that occurs in the normal cell.

During transformation, dedifferentiation is not uncommon, conse-quently, the cell-specific isoforms of many cytoskeletal proteins may not be found. The class of intermediate filament protein expressed in the original tissue, however, appears to persist, although vimentin ex-pression may resume with transformation. This fidelity of intermediate filament expression may allow identification of metastases [115].

Degrees of transformation can be distinguished both morphologically and by their biochemical response to substrate and shape. These will be discussed in more detail later. The variations in cytoskeletal pattern suggest that the transformations produced by viruses, tumor promoters, or other agents may either progressively modify one filament system or involve additional filament systems. There have been no systematic examinations to distinguish between these two possibilities.

3.7 Protists

The eucaryotic cytoskeleton predates the animal cell. Just as muscle

45

and epithelia can be viewed as variations on themes already expressed in fibroblasts, so animal cells in general can be viewed as variations of structures that have been seen in the unicellular protists.

Microtubules, microfilaments, and their slender cross-bridges are widely distributed among the protists. For example, in forams the uptake of food particles is mediated through a branched work that includes microtubules and slender 5 nm filaments which do not decorate with myosin [116]. This anastomosing network is particularly suitable for studying saltation, discussed in detail later.

Various amoeboid cells have contributed to studies of amoeboid locomotion. Because protists are easily grown, large amounts of material permit purifying proteins for biochemical studies. In addition, some of them, such as *Physarum*, provide useful morphological subjects to study contraction and relaxation.

During the early stages of tension, *Physarum* contains parallel bundles of microfilaments, crosslinked by short bridges. As tension increases, dense areas that may represent thick filaments of myosin appear between the microfilaments. As the cell relaxes, the dense areas disappear and the microfilaments form less organized bundles. Related reorganizations are visible during isotonic contraction as well, but at no time during these cycles does the density of the microfilaments change radically; it has been suggested that the regulation of these states is principally modified by the assembly state of the myosin [117].

Naegleria gruberi is a particularly striking protist since it undergoes a carefully regulated transition from an amoeboid cell that walks along surfaces to a slender, streamlined, flagellated cell that swims. This transition is regulated by simple physiological cues and can occur synchronously. Before the transformation *Naegleria* appears to use an actin-based motility system, although microtubules are present during mitosis. After transformation, a subcortical cage of microtubules generates the cell shape, as well as the usual 9 + 2 axonemal structure in the two flagella. Thus *Naegleria* is an elegant system for monitoring the transition between these two systems and the regulation of gene expression for cytoskeletal proteins [118].

Tetrahymena is a permanently ciliated cell, unlike *Naegleria*. The subtle and flexible cell shape of *Tetrahymena* appears to be generated by a meshwork that underlies the plasma membrane and includes microtubules. This meshwork also contains material antigenically related to intermediate filament proteins. The control of cell shape and local patterns in *Tetrahymena* is particularly informative for what it tells us about local and global controls over shape and pattern, and will be discussed later.

Chlamydomonas has been fruitful, because it is well characterized ultrastructurally and genetically. It is not a fastidious organism, and can be grown in large amounts. *Chlamydomonas* also has both microfilament and microtubule systems. The most conspicuous microfilament

46

system in *Chlamydomonas* is the projection of microfilaments that forms during mating. A slender process extends from the apex of the cell to connect one mating type to the other; this process contains a bundle of thin filaments that decorate with heavy meromyosin. It is likely that actin plays other roles, as yet undocumented, in the life of *Chlamydomonas*. Microtubules form a cage beneath the cell membrane that radiates from the pair of basal bodies underneath the flagella at the cell apex. Microtubules also extend from the centrioles of the basal bodies in a 9 + 2 axonemal structure. *Chlamydomonas* is the major source of understanding of axoneme assembly. Numerous motility mutants have been identified and mapped, and these have been exploited by combining conventional complementation tests and *in vivo* complementation using radiolabeled cells. The axoneme is a complex structure, containing over 50 distinct polypeptides; its assembly is probably sequential, like that of bacteriophage. For many of these proteins, no functional flagellum is formed in their absence; for some of them, a series of other proteins is blocked from entering the structure until the mutant protein has been replaced by a functional one. Possibly most striking is that one of these keystone proteins changes in efficiency depending on its phosphorylation [119]. Flagellar regeneration in *Chlamydomonas* has also been a rich source of information about the regulation of tubulin genes [120].

Biochemical and morphological evidence show that microfilaments and microtubules are widely distributed among protists. The full extent of their distribution and the full range of their various specializations have scarcely been touched. It is clear, however, that both systems predate the metazoa, and they are probably of universal distribution. Intermediate filaments are far less conspicuous among the protists. There is at present one case of antigenically related proteins in *Tetrahymena*, and one in *Candida* [121]. This reduced visibility may be because the major metazoan role for intermediate filaments is the integration of cells into tissues, seen especially in the roles of the cytokeratins and desmin. However, the juxtaposition of intermediate filaments and microtubules, and the apparent relationship between intermediate filaments and nuclear location suggest other possible roles for intermediate filaments. Thus, the possibility remains that some precursor to intermediate filaments will be found as widely distributed among the protists as actin and tubulin.

3.8 Plants

Plant development differs greatly from that of animals. It involves multiplication and enlargement *in situ*, with no cell rearrangements, and the importance of the wall is so great that it might first seem surprising that the cytoskeleton is as central to plant development and maintenance as it is to animals. However, despite less attention given to and the experimental obstacles presented by plant cells, it is clear that both actin- and tubulin-based sytems are central to the organization of plant cytoplasm [122].

Actin in plants is known principally by its connection with cytoplasmic streaming, mediated through cables anchored in the plasma membrane. These cables of actin consist of bundles of microfilaments all with the same polarity, i.e. with the barbed end anchored in the membrane. The precise mechanism by which streaming proceeds is not understood in detail, but also appears to involve myosin. Actin may well play other roles, as yet unknown.

Microtubules have been the central focus of cytoskeletal studies in plants, possibly because they are more readily visible by electron microscopy. The advent of suitable immunofluorescence techniques will extend the range of structures examined in plant cells [123]. Microtubules change location at each stage of the cell cycle. During interphase, many microtubules lie in a cortical region underneath the plasma membrane, freqeuently parallel to the cellulose microfibrils that overlie the microtubules within the cell. Cross-bridges between the microtubules and the plasma membrane have frequently been seen, but the mechanism by which microtubules orient cellulose fibrils is not yet understood [124].

Before prophase, the preprophase band appears as a wide band of microtubules somewhat more dense than the neighboring microtubules elsewhere in the cortex. The microtubules continue to pack together and other cortical microtubules are reduced as the preprophase band matures. The nuclear envelope, which has no associated microtubules during interphase, begins to show signs of tubulin as the preprophase band develops. Microtubules arranged both radially and as a meshwork occur in a region between the nucleus and the cell cortex. Some of these microtubules appear to connect the preprophase band and the nuclear envelope. The preprophase band then disappears before prophase itself. A striking property of the preprophase band is that it presages the plane of cell division after nuclear division [125].

At prophase, most of the microtubules are found in a cage around the nuclear material. As mitosis progresses, the spindle becomes organized and the chromosomes proceed through metaphase and anaphase. The principal difference in appearance between the plant spindle and animal spindles is that, in the absence of centrioles, the plant spindle is more barrel shaped than biconical.

At telophase, the microtubular content of the spindle reduces and microtubules begin to appear in the phragmoplast, which participates in cell division by orienting the formation of the cell plate. The phragmoplast microtubules appear first in the center of the cell and begin to extend laterally until the entire plate is formed.

Following cell division, microtubules gradually take up their interphase cortical situation. If microtubules play a role in the orientation of cellulose fibrils, they generate the physical constraints on the future

48

shape of the cell. Between determining cell shape during interphase and constraining polar divisions of cells via the preprophase band, microtubules are central to plant development. It is likely that there is a role for actin in all of this, but until now technical constraints have made it difficult to determine what that role may be.

Eucaryotic cells organize their intracellular and extracellular activities by specific modifications of the cytoskeleton. In most cases, cellular specialization is reflected in cytoskeletal specialization. This universality presents both advantages and hazards. On the one hand, it gives hope that important mechanisms may be widely distributed and only slightly modified among different eucaryotic cells. On the other hand, the highly specialized forms frequently chosen for study are precisely that: constellations of proteins which have been modified for either extreme values of elasticity or contractility or for greater stability than in the majority of cells. Such specialized cases have been helpful in the past and will continue to be so, as long as the functional constraints of the system are also carefully weighed.

4   Cytoskeletal choreography

4.1 Cytoskeletal drugs and drugs with secondary cytoskeletal effects

For both the microfilament and the microtubule systems, drugs exist that specifically bind to tubulin or actin in either the monomer or polymer state, and are effective *in vitro*. There are numerous parallels between the various cytoskeletal drugs. Thus, for both microfilaments and microtubules, there are drugs that bind to the monomer and poison further assembly of the polymer; conversely, there are drugs that bind to the polymer and increase its stability.

Antimitotic drugs were the first cytoskeletal drugs known; colchicine is the oldest of these. Colchicine, nocodazole or oncodazole, and colcemid bind at a common binding site on the tubulin dimer [126]. Binding of the drug increases the nucleotide triphosphatase activity of the dimer. The drug–dimer complex can bind to microtubules at the fast assembly end, and decrease the rate of further addition at that end. Consequently, the critical concentration of the solution is raised to that of the slow assembly end. Colchicine dissociates slowly and exhibits non-specific binding to cell membranes, in particular, to the nucleoside transporter. Nocodazole has the advantages of increased specificity and more rapid reversibility. One control for side effects of colchicine is lumicolchicine, whose binding to the tubulin dimer is inactivated by light but exhibits the same pattern of non-specific binding. Part of the binding site occupied by colchicine is shared by podophyllotoxin with similar effects. A different binding site is occupied by vinblastine and vincristine. When vinblastine binds to the tubulin dimer, it causes aggregation and crystal formation; it decreases polymerization by recruiting monomers into these non-tubular aggregates. The net effect is to lower the concentration of free tubulin monomer reversibly. Overall, the effect of all of these drugs is to decrease the rate of polymerization, not to disassemble intact microtubules directly. Their effects on cytoplasmic microtubules take one to several hours to be seen, although they prevent the formation of the mitotic spindle immediately. Because these drugs act primarily by poisoning the growing polymer, they exhibit highly specific effects on different classes of microtubules. Flagellar microtubules are unaffected, once assembled, while mitotic microtubules never form.

Taxol, by contrast, is a drug which binds stoichiometrically, not substoichiometrically, and which stabilizes the microtubule. There is little if any binding to the tubulin monomer outside of the microtubule.

Overall, taxol increases the stability of microtubules and lowers the free concentration of tubulin. It abolishes the lag phase during polymerization; all its effects are reversible [127].

Cytochalasins are a group of drugs that bind reversibly to actin monomers and in binding increase the nucleoside triphosphatase activity of actin. The actin—cytochalasin complex can bind to the fast assembly end of the microfilament and prevent further polymerization from that end [128]. It thus raises the critical concentration to that seen from the slow end. The cytochalasins can eliminate the lag phase during actin polymerization, presumably by increasing the stability of a dimer or trimer. However, growth from such nuclei occurs with the kinetics expected from the slow assembly end. Cytochalasin B has the drawback of inhibiting glucose transport at the cell membrane; however dihydro-cytochalasin B and cytochalasin D do not affect glucose transport [129]. Cytochalasin D has the higher affinity for actin. Just as taxol stabilizes microtubules, so the phalloidin drugs stabilize microfilaments. Phalloidin and phallacidin bind to polymerized actin stoichiometrically and lower the critical concentration reversibly.

Thus, for both microfilaments and microtubules, drugs are available that specifically block polymerization and that specifically block depolymerization. Drugs are not as yet available which cause the prompt depolymerization of either system, with the possible exception of maytansine which *in vitro* depolymerizes microtubules rapidly. Maytansine is however an alkylating agent; its anti-microtubule effects are somewhat irreversible and perhaps non-specific.

In using all the cytoskeletal drugs it is important to distinguish prompt and delayed effects of the drugs. Thus taxol and phalloidin promptly stabilize their respective filaments; a delayed effect within a living cell is a bundling together of microtubules or microfilaments respectively. This delayed effect of bundling is inhibited by inhibitors of metabolism, suggesting that it is energy dependent. Bundling is also not seen in cytoskeletons treated with the drugs. The microtubule poisons promptly inhibit mitosis, but require one to several hours to eliminate cytoplasmic microtubules. Again this depolymerization of the cytoplasmic microtubules is inhibited by metabolic inhibitors suggesting that depolymerization is energy dependent as well. Colchicine does not disassemble microtubules in cytoskeletons suggesting that disassembly of microtubules in the living cell is a dynamic process, involving an energy-dependent regulation of the ends of microtubules. The cytochalasins promptly inhibit ruffling and filopodial extension. However, extended periods of time are required for the formation of dense foci and arborized projections from the cell. At the ultrastructural level, there are prompt and delayed effects of cytochalasins. Microfilaments appear to have free ends immediately after drug treatment, but take time to form clusters and bundles [130]. Cytoskeletons treated with cytochalasins display more free ends. However, the cytoskeletons do not undergo the clumping seen in the living cell. This clumping appears to be an energy-dependent reorganization that is subsequent to the primary site of action of the cytochalasins.

Unfortunately, there are no small metabolites that directly affect intermediate filaments. At present, the only means of disassembling intermediate filaments *in vivo* is by microinjection of certain antibodies. In addition, microtubule poisons will lead, over time, to the formation of vimentin, and in some cases desmin caps, but this is a delayed indirect effect of these drugs, since the intermediate filament coils are only beginning to form when microtubules are completely disassembled.

Numerous drugs whose primary sites of action are known have marked secondary effects on the cytoskeleton and its organization. Thus, inhibitors of energy metabolism, such as DNP, lead to the loss of stress fibers in about an hour and a half [131]. They also prevent microtubules and microfilaments from bundling in taxol and phalloidin. Inhibitors of energy metabolism inhibit the depolymerization of microtubules by the microtubule poisons and in general prevent any energy-dependent step of cytoskeletal organization or reorganization (*v.i.*). Inhibitors of protein synthesis unexpectedly cause the formation of juxtanuclear vimentin caps [132]. Thus any agent which causes protein synthesis inhibition, either a direct inhibitor or an agent such as interferon or heat shock, can lead to cytoskeletal rearrangements. Cyclic AMP and its related agents, by activating cyclic AMP-dependent protein kinase, will affect the phosphorylation of myosin light chain in non-muscle cells and the phosphorylation of MAP proteins. cAMP may affect muscle cells by altering the phosphorylation of desmin [133]. Several calmodulin-binding proteins are known in the cytoskeleton, among them spectrin, so that anti-calmodulin agents can act at several loci in the cytoskeleton. The calcium-stimulated disassembly of microtubules in extracted cytoskeletons is inhibited by calmodulin inhibitors. This potentially calmodulin-mediated depolymerization of microtubules is counteracted by MAP proteins. Sodium butyrate, an agent used to inhibit histone deacetylation, also has numerous effects on the cytoskeleton, some of them occurring within several hours [134]. Whether these result from inhibiting a cytoskeletal deacetylase or substituting for acetyl groups in some way is not known. Drugs such as verapamil that inhibit calcium uptake by the cell clearly have a wide range of potential cytoskeletal effects.

Finally, as will become clear in later discussions, none of these filament systems is wholly independent of the other two. Thus, the delayed effects of any given drug may be indirect and mediated by a different filament system than the one in question.

4.2 Control of cell movement and shape
To describe the full range of cell locomotion and shape in eucaryotic

cells would require several books. However, several important aspects of locomotion and shape emerge from considering three major classes of cell movement. These three forms of locomotion will make clear in what ways cell shape can be a function of cell movement.

Many cells move by swimming, i.e. by movements that depend upon cilia or flagella, peninsular appendages that contain an axoneme based on a 9 + 2 arrangement of microtubules. The 9 + 2 axoneme is widely distributed among eucaryotes and is largely conserved in its morphology. Despite similarities in ultrastructure, however, the axoneme executes various patterns of movement, depending upon the accessory proteins of the particular cilium or flagellum. Some cells, like *Chlamydomonas*, use a breast stroke motion in which the power stroke is a nearly straight beat pushing backward, with a recovery stroke in which the flagellum is curled to present a minimum surface to the fluid. Other cells beat the flagella behind to propel themselves forward. Still others, like some classes of ciliates, exhibit large numbers of short cilia which can be used to swim or as legs to walk on a surface. In all these cases, the motile force is generated by the ATPase, dynein, that can cross-bridge between the outer doublet microtubules of the axoneme reversibly, with a power stroke step involved at some point that generates sliding of one doublet pair relative to the other. The additional geometric constraints placed on the axoneme by the accessory proteins then determine the pattern in which this dynein-generated force is exerted [135].

Most cells that swim have a cell shape that appears to result from microtubules, frequently subcortical, lying a small distance underneath the cell membrane. Such submembranous networks generally propagate from one locus within the cell, frequently from paired basal bodies. However, the mechanisms by which detailed cell shapes are produced, ones that frequently involve asymmetric shapes or solids of revolution, is not yet well understood.

Many cells, such as amoebae and macrophages, engage in amoeboid motion. An amoeboid cell viewed from the side has several discrete domains in the cytoplasm (see Fig. 4.1). The ectoplasmic region is a peripheral domain of the cytoplasm that lies just under the cell membrane. Deeper, in the domain called the endoplasm, the cytoplasm is birefringent and visco-elastic. At the front of the cell is the pseudopod tip, with a transition between the endoplasm and the ectoplasm believed to result from contraction. Since in this domain the endoplasm and ectoplasm appear to be in a continuous, molecularly interconnected gel, the contraction at the periphery of the frontal ectoplasm pulls up into it the gelled endoplasm Direct observations confirm that the ectoplasm and endoplasm are visoelastic solids [136, 137]. The only domain of an amoeboid cell with a low viscosity is the zone of shear between the endoplasm and ectoplasm, and a small region in the tail region where material from the

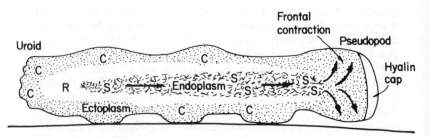

C = Contracting   R = Relaxed   S = Stabilized

Fig. 4.1

ectoplasm is recruited into the endoplasm by a transient solation of the ectoplasm material.

These behaviours appear to be regulated predominantly by the concentrations of free calcium and ATP. In domains with ATP and abundant calcium, contraction occurs. In domains with abundant ATP but little calcium, there is relaxation; in the endoplasmic interior there appears to be limited amounts of ATP and calcium [137]. Finally, this regional activity results in locomotion if some parts of the cell are attached to a substrate. Without attachment, the cell can fluctuate in shape without net locomotion.

In this mechanism, there is no role for microtubules; many amoeboid cells appear to be insensitive to colchicine [138]. Secondly, the shapes of amoeboid cells result from local contractions. Amoeboid motion is sensitive to cytochalasin B. At a fairly low concentration, 10^{-7} M, cytochalasin B can inhibit macrophage locomotion [139]. Moreover, it inhibits locomotion substantially more rapidly than the decrease in average filament length occurs when cytochalasin is added to microfilaments *in vitro.* Cytochalasins may inhibit amoeboid motion by competing with some cellular protein for the fast assembly ends of microfilaments.

Cell movement in amoeboid cells appears to be controlled at the cell membrane. The leading edge of a leukocyte is more responsive to chemotactic factors; the cell appears to steer towards chemotactic signals. Steering does not require a nucleus or microtubules [140, 141]. Anucleate fragments of leukocytes which contain neither nucleus, centriole, nor microtubules respond chemotactically. In leukocytes, the microtubules during interphase may anchor and orient cellular structures, such as the nucleus and cytoplasmic granules.

A third major class of cellular locomotion may be called microtubule dependent or fibroblastic movement. If a cell engaging in this form of locomotion is observed from the spherical state that follows mitosis or replating in culture, three stages can be described: initial spreading, polarization and translocation. This appears to be an obligate sequence; a cell that cannot spread cannot polarize, and a cell that cannot polarize cannot move [142].

The architecture of spreading fibroblasts and epithelial cells has been described. During spreading the cell extends filopodia, ruffles and blebs, all of which can adhere to the substrate or other cells and generate force. If such adhesions have formed with a substrate, a cell is spread after a number of minutes. If such contacts engage detachable particles, the cell will clear out a circular area. This process depends upon extracellular K, Na, Cl, and either Mg or Ca at a neutral pH. In addition, it is energy dependent. The forms of extensions are affected by the conditions used for washing and plating cells. All three forms of extensions and particle clearance are inhibited by 1 μg ml^{-1} cytochalasin B, but insensitive to antimicrotubule drugs. This sensitivity seems reasonable, since all three structures contain microfilaments but not microtubules [143].

Cells that clear an area around themselves in this manner are generally circular. Deviation from circularity requires an additional process, polarization, which can be inhibited by antimicrotubule drugs; polarized cells reverse their morphology after the addition of such drugs. Cells prevented from polarizing by colchicine have an additional defect; vimentin or desmin filaments slowly form a dense cap near the nucleus (v.s.). When such caps are formed by adding an antibody to intermediate filaments, cells do not lose their polarization [144]. Not surprisingly, the microtubule organizing center (MTOC) is oriented in the direction of cell movement. When cells are induced to walk in the same direction by wounding a confluent monolayer, the cells move into the wound and MTOCs are found toward the leading edge of the cell. Drug studies suggest that MTOC rearrangement is a precondition for cell movement. In cells inhibited from moving with cytochalasin B, MTOCs will reorient; however, if MTOC reorientation is blocked by colcemid, no cell movement occurs (145).

The control of cell movement in microtubule-dependent locomotion is not yet fully understood, but several correlations appear significant. Large numbers of cells can be followed by tracks left in a field of gold particles. Such techniques show that the main actin-containing bundles appear to parallel movement. In addition, the MTOC and, if present, the primary cilium point in the direction of movement [146]. Although stress fibers are oriented in the direction of movement, they are present in inverse proportion to the velocity (v.s.).

The pattern of cell movement presents several features that must be explained by a complete model. Each cell line has a characteristic pattern of cell motility, i.e. an average turning frequency and velocity; no cell line exhibits a deterministic pattern [147]. Secondly, sister cells often resemble each other or their parents in the pattern in which they move. About 60% of 3T3 cells display identical or mirror image tracks when compared to their sisters. Finally, a moving cell that collides with another cell will frequently move away from the collision at an angle symmetrical to the original impact.

Clearly, some elements of cell movement can be transmitted through mitosis, in which cells round up, while other aspects reflect cell interactions with its environment. Several environmental factors have been identified. Fibronectin (also called LETS or CSP) mediates attachment of fibroblast-like cells to their sub-stratum, in particular to collagen. Fibronectin added to the medium increases the spreading of transformed cells and motility of both normal and transformed cells [148]. Another major extracellular factor is calcium concentration, since most cells examined have an absolute requirement for extracellular calcium for motility [149]. It is not clear why calcium is required, since the exploratory movements of filopodia and ruffles are not inhibited in its absence.

One approach to determining cellular control over the pattern of locomotion is to ablate specific structures and determine whether the remaining structures can still move. Both microtubules and microfilaments are required in the microtubule-dependent cells (*v.s.*). Intermediate filaments in some cells can be effectively removed from the cytoplasm by microinjection of a vimentin-specific antibody. The vimentin filaments form a cap that persists for over 24 hours. Such cells nonetheless have a normal shape and can move across the substrate. Intracellular organelles remain in the appropriate locations and the cell proceeds with mitosis and cytokinesis [144]. It is not known whether cells with such vimentin caps can still express the sister cell similarities seen in intact cells. Cells do not require nuclei for many of these behaviors. Fibroblastic and epithelial cells enucleated with cytochalasin can attach, spread, form a characteristic shape for that cell line, engage in pinocytosis, and move in appropriate ways, including contact inhibition of movement [150]. Again it is not known if such enucleated cells retain the detailed memories exhibited by sister cells. The extent of ablation that is required to seriously restrict the spatial repertoire of cells or cell fragments is surprising. Selective disruption of cells following cytochalasin produces small cell fragments, one to a few microns across, called microplasts. These microplasts engage in extremely stereotyped motile behavior. One microplast will repeatedly ruffle, or bleb, or extend filopodia. These behaviors persist over a period of hours without changing. Microplasts recoil from contact with another cell, but they engage in no other spatial behaviors. This suggests that for the cell to co-ordinate motion, something integrates these various restricted behaviors.

These patterns of motility and shape changes are not a peculiarity of cultured cells. Observations of embryos show that cells *in situ* undergo comparable shape changes during locomotion, and that many aspects of motility observed in culture represent the naturalistic behavior of cells. The movement of cells up adhesive gradients, a property easily demonstrated in tissue culture cells which results from increased frequency of attachment during exploration, has been proposed as a major determinant of embryonic morphogenesis [151]. If so, cell motility and its controls are one of the major determinants of animal development.

Cell shape in cells that use this interactive form of movement does not resemble either the deterministic shape of swimming cells, nor the completely dynamic shapes of amoeboid cells. Rather, in cells using interactive motility the various filament systems interact at all times and produce both cell movement and the moment-to-moment cell shape. This interaction is visible at the ultrastructural level. All three major filaments have been observed in contact with each other [64]. Structural integration is also suggested by behavioral co-ordination. When the trailing tail fiber of moving cells detaches from the surface, the ruffling activity at the cell front increases. The tail contraction has two components, a fast energy-independent component, probably elastic, and a slow component that requires energy and is probably contractile.

In some cell types, interaction between the filament systems is particularly accessible to experimentation. Neuroblastoma cells extend neurites which branch and radiate away from the cell body. Neuroblastoma cells also display some memory for cell shape. After mitosis, the extension of neurites can be blocked by either nocodazole or cytochalasin. Thus, both microfilaments and microtubules play a role in neurite extension. However, the microfilament involvement in neurite extension may be in motility, because extended neurites do not retract when treated with cytochalasins. In contrast, extended neurites retract when exposed to nocodazole. After retraction, the cell has some memory of the shape it displayed previously, since removel of the drug leads to re-expression of the previous pattern. Finally, neurite retraction is co-operative; if the cell is treated simultaneously with nocodazole and cytochalasin, neurites do not retract. Moreover such neurites still contain intermediate filaments which normally recoil into the cell center when neurites retract [152]. This suggests that the intermediate filament coiling observed with antimicrotubule drugs is mediated by interactions between intermediate filaments and microfilaments. Microtubules cannot extend neurites in the absence of microfilaments; depletion of microtubules cannot cause neurites to retract unless there are intact microfilaments. Intermediate filaments are clearly interacting with both microtubules and microfilaments. Comparable interconnections are seen between the three filament systems in many cell types.

In other cell types, it is possible to identify several contributions to cell shape from a single filament class. When platelets undergo the shape change described above, various actin-associated proteins have different locations. At this time, the association of actin with the skeleton changes. After activation, actin is found in a cytoskeletal core with much of the myosin and actin-binding protein of the platelet. Morphologically, the myosin and actin-binding proteins occupy different domains within the platelet. These distinct associations can be inhibited differentially. If platelets are treated with cytochalasins before thrombin activation, the myosin—actin gel forms normally, but actin-binding protein does not associate with the cytoskeletal core and filopodia do not form. In contrast, phorbol myristate acetate (PMA) activation promotes filopodial formation but the cytoskeletal cores do not contain myosin. Architecturally distinct regions behave in functionally distinct ways, yet platelet activation normally involves the co-ordinated interaction of both aspects of the actin-based system at once [153].

It is likely that land plants also have an interactive mechanism for determining cell shape, while microtubules play a major role in modeling the cellulose fibrils of the wall during cell growth. Plant cells exhibit a wide spectrum of shapes that are developmentally regulated, unlike many protists which have deterministic cytoskeletons. However, the role of actin or other filament systems in this modeling is not yet understood.

4.3 Cytoskeletal interactions with the plasma membrane and the extracellular matrix

The cytoskeleton makes numerous contacts with the cell membrane; presumably some membrane proteins interact with cytoskeletal elements. In the simplest model for a membrane, the isolated lipid bilayer has membrane proteins floating in it like icebergs in a lipid sea. If this were an adequate model, the mobility of both lipids and membranes could be predicted by physical chemistry. When cells are examined, however, only lipids behave as predicted. A substantial fraction, from 10—80%, of the membrane protein exhibits no lateral mobility; proteins which do move generally move 10—100 times slower than predicted. Lateral mobility of membrane proteins is raised by treatments that release the plasma membrane from its cytoskeletal association; thus, many membrane proteins interact with the cytoskeleton, either directly or by associations within the membrane to other proteins that interact directly [154]. Such membrane—cytoskeleton interactions can change during development. In the youngest muscle cells, acetylcholine receptors are freely mobile. At later stages, an immobile fraction increases with time in patches and large clusters. The extremely mobile fraction is readily extracted by detergent, while the immobile fraction remains associated with the cytoskeleton after the removal of soluble proteins and lipids [155].

Organized interactions between both microfilaments and microtubule systems and membrane sites have been seen. Most actin filaments associate with the membrane by their fast assembly end. Apparent

interaction between microtubules and the membrane is seen in cyto-
toxic lymphocytes. These killer cells orient towards a target so that the
MTOC faces the intercellular contact, although other conspicuous
membrane structures such as antibody caps are not organized in any
particular relationship to the MTOC [156].

The cell can generate or maintain an asymmetric distribution of
many membrane components. One such phenomenon, capping, has
been explored in detail. Many membrane proteins are dispersed in the
absence of ligand and rearrange after binding ligand. If an initially
symmetrical cell is exposed to a ligand at 4°, the uniform distribution
of receptors is maintained. If such a uniformly labeled cell is then
warmed up to 20° or 37°, the dispersed receptors begin to aggregate
into patches. Patching does not occur in mitotic cells, and can be
prevented by high doses of multivalent lectins such as Con A that bind
cell membrane proteins [157]. Depending upon the ligand and receptor, two routes are open to these patches. Some multivalent ligands in-
duce a cap to form passively in the presence or absence of energy inhibitors, microfilaments or microtubules. Passive caps require half an
hour to an hour to form, and cells on which they form usually remain
symmetrical in shape.

When receptor—ligand complexes rapidly form caps, it requires
energy and extracellular calcium. Such caps form on cells that are no
longer spherically symmetrical. The protuberance on which the ligand—
receptor complexes congregate is underlain by actin, myosin, calmodulin, α actinin and myosin light chain kinase [158]. The myosin
in the cap appears to have phosphorylated light chains and would be
more competent for contraction (*v.s.*). Rapid capping is inhibited by
cytochalasins but not by antimicrotubule drugs.

A simple explanation for these phenomena is that in the actively
capping cell, ligand—receptor complexes are pulled by a contractile
mechanism into the region of the cap. In other asymmetric cells, cells
undergoing phagocytosis, chemotaxis, or cleavage, domains of high
microfilament density have ligand—receptor complexes aggregated over
the filament densities. The filament density is not enough in itself to
cause such aggregation, since in a chemotactic cell there is also high filament density at the cell anterior which does not collect ligand—receptor
complexes. A model for these diverse phenomena proposes that a local
deformation of the membrane, or wave, moves over the cell surface.
Properties of such a membrane wave, such as microviscosity and surface
tension, would be likely to differ from those of the bulk membrane. The
model proposes that membrane proteins interact with such a wave
differently, i.e. some are unaffected by it, while others are entrapped
and move with the wave towards regions of high microfilament density;
there they form associations with the submembranous microfilaments.
Waves have been seen on cell surfaces; since these waves are probably
generated by microfilament contractions, the model explains why such

membrane rearrangements are sensitive to cytochalasins and inhibitors of energy metabolism [159]. It also suggests a mechanism by which lipid components could rearrange.

Whichever model is correct, a protein's distribution in the membrane can be altered by its binding ligand, and this change in distribution is in some way mediated by the cytoskeleton via an energy-dependent mechanism. Furthermore, such rearrangements are frequently co-ordinated, in that several membrane components may be rearranged by the interactions of one of them. Finally, the range of rearrangements possible is a function of cell state, and is severely restricted in mitotic cells or cells in which a significant fraction of the surface membrane proteins are inhibited in their mobility [157, 159].

These surface movements play several roles. By changing the distribution in the membrane, they can change the apparent affinity of surface receptors. By causing rearrangements of the skeleton, they generate potential signals to the interior. Rearrangement is required as the first step in internalization and presumably plays a role in forming attachments to the surface. A comparable mechanism is probably responsible for the movement of particles over the cell surface before internalization. Many of these roles are important to cell movement during embryogenesis and during wound healing; they may also be related to the role played by the extracellular matrix (ECM) in regulating gene expression. Tissue interactions mediate much of embryonic development and require contact between cells or between cells and extracellular materials in most cases. Clearly a solid, external substrate that interacts with cell surface proteins can generate patterns within the cell by those interactions. For many cell types, appropriate differentiation can occur only in the presence of the specific ECM normally associated with it [160]. The continued expression of specific differentiation products may require continued exposure to a specific ECM. The mechanism by which the matrix instructs the cell is not yet understood, but some aspects of gene expression and the cytoskeleton will be discussed in detail later.

4.4 Intracellular movement and transcytosis

Most cell structures have either a characteristic location or pathway through the cell. These myriad movements can generally be assigned to one of two categories. Within the central region of the cell, primarily occupied by microtubules and intermediate filaments, are found a large range of intracellular organelles. Some, such as the Golgi apparatus and lipid droplets, are found anchored at a fixed location, although both of these organelles can undergo specific relocations. In contrast, the endoplasmic reticulum (ER), lysosomes, and mitochondria are deformable and to some extent motile. The ER is limited in its motility, but both lysosomes and mitochondria engage in various motions within the cell. In addition, special high contrast microscopy reveals smaller intracellular organelles which move in the central domain [161].

These central movements depend on intact microtubules, are generally resistant to cytochalasins, and are independent of each other. Two particles in close proximity are unlikely to display precisely the same pattern of movement [162]. Microinjection of vanadate does not inhibit these movements [163], but vanadate does inhibit such movements in permeabilized cells [164]. Therefore, the mechanochemical proteins involved are not identical to flagellar dynein, but may be related. In addition, these movements are intermittent; that is, any particle moves steadily for a brief period and then abruptly halts. This pattern of intermittent, regular motion, saltation, can be distinguished from Brownian motion by its kinetics [165]. Significantly, particles not engaged in saltation are fixed in position. Moving particles are often seen in contact with microtubules, suggesting that the microtubule dependence is fairly direct. The only selective inhibitor of central saltatory motion is treatment with a hypertonic medium [162], which does not stop ruffling. Saltation is seen in plants, during cell plate formation [166].

Saltation appears to have an elastic component. Individual particles that arrest occasionally engage in abruptly faster motion forward, as if retracted by some matrix component [165]. This elastic component is most clearly suggested in systems in which cytoplasmic particles undergo cycles of movement [167]. Chromatophores contain pigments of various colors whose distribution is under the control of the cell. The particle distribution changes between central aggregation and uniform dispersal throughout the cell. The outward motion is saltatory and energy dependent. Appropriate stimuli cause a coherent, unidirectional inward movement that is energy independent. In some chromatophores, associations between pigment granules and the cytoskeleton are readily seen in high voltage electron micrographs (HVEM). These connections are an irregular lattice of variable dimensions, whose components are called microtrabeculae. The uniform and cell-wide inward movement of pigment granules suggests that in these cells the elastic component has been specialized and elaborated beyond that normally seen in other cells.

Similar intracellular movement has been observed in a wide range of cells, including plants, protists and animals. Axoplasm squeezed out into a carefully chosen buffer provides an *in vitro* model. Using this axoplasmic system, many properties of cortical movement have also been determined. The structures that move in the cortical , microfilament-rich region are generally small, and include clathrin-coated vesicles, synaptic vesicles and secretion vesicles. In this region, all the detectable vesicles either move or are anchored. Cortical and central motion require energy but not calcium. Cortical movement also has an elastic component and is cytochalasin resistant, despite the predominance of microfilaments in this region. Cortical movements are predominantly unidirectional and constant, without the intermissions seen in the central region. Since this movement occurs without the plasma membrane, clearly neither membrane proteins nor an action

61

potential are required. However, like the central region movements, it does require oriented filaments; if axoplasm is disordered by stirring, the movement continues in random patterns [168].

Some intracellular movements involve a complete transition between these domains and through the cell membrane, and exist in two forms. In the outward direction, exocytosis includes secretory processes in which secretion granules form, move to the cell surface, and release their contents to the outside. In the opposite direction, endocytosis consists of several internalization processes. Secretion requires many steps, beginning with synthesis of the protein on the rough ER, passage through the Golgi into secretion granules, and movement to the cell surface. Release of material at the surface may either be continuous or under the control of a stimulus. Movement of material from the rough ER to the Golgi appears to be organized by microtubules, which are required for a coherent process of transport [169]. This interaction between microtubules and the rough ER is further confirmed by the observation that taxol induces large complexes of microtubules with rough ER [170]. These complexes probably represent a sink for the excess microtubules formed by taxol; the cross-bridges visible in them may represent ones present in the untreated cell but normally fewer in number. The final stages of secretion under a specific stimulus do not appear to require microtubules. At this point, vesicles in the peripheral cytoplasm are largely associated with microfilaments, which may play a restraining role in preventing the vesicle from touching the cell membrane, because cytochalasins stimulate secretion in some cases [171].

Movement through the cell from the exterior to the interior occurs via three paths. Phagocytosis occurs when a cell engulfs an object by spreading its membrane around the particle [159]. Engulfment is frequently accompanied by numerous microfilament associations at the point of phagocytosis. When the membrane stretches completely around the particle fusion occurs. Phagocytosis is sensitive to cytochalasin. During pinocytosis, 200–700 nm vesicles form and engulf a volume of the extracellular fluid. The amount of pinocytosis depends on the cell type and allows fluid phase nutrients or signals to be taken up. Pinocytosis is also sensitive to cytochalasins [171]. The third form of internalization will be called endocytosis. Endocytosis has also been used to name all three processes, but is now generally restricted to one in particular, the one by which 60–70 nm vesicles form that are coated with clathrin. These vesicles transport internalized material to the lysosome; at least the initial stages of this internalization are resistant to cytochalasins, although it is not clear whether the transition to the lysosome is prevented [171]. All three processes are often cyclical, and can occur at high frequencies. The cell therefore has a mechanism by which internalized membrane can return to the plasma membrane.

The last class of intracellular movements is that of the cytoskeletal elements themselves. In neurons, the cytoskeleton is laid out in an elongated, linear array that allows analysis of movement rates. In these cells, two waves of cytoskeletal migration from the cell body, called the slow components a and b, are clearly discernible. SCa comprises microtubules and neurofilaments. SCb includes actin and other proteins. Both components include proteins thought of as cytoskeletal and also proteins usually thought of as soluble, such as enolase and creatine kinase [22]. That these enzymes move with the slow movement of the cytoskeleton itself suggests that enzyme associations to cytoskeletal elements may be general [22].

How are intracellular movements regulated? Interphase micro-tubules are predominantly of a single orientation, with fast assembly ends towards the cell periphery (v.s.). Therefore, models in which antiparallel microtubules are required as railroad tracks for opposite directions are impossible. Perhaps the simplest model would be one in which particles can either engage or not to the microtubule system, but are always capable of interacting with the elastic component found in between the microtubules. Thus if a particle engaged the microtubules, it would head outward; when it disengaged it would either rest or be elastically drawn back into the central part of the cytoplasm. Unfortunately, there is no comparably simple model available for the basis of movements within the cortical region. It is possible that these move-ments are kinetic, rather than tactic, and are oriented in the axon by a specialized mechanism.

4.5 Mitosis and assembly

Mitosis presents perhaps the single largest challenge to understanding spatial order in eucaryotic cells. In the span of a few minutes, major reorganizations of the chromosomal material occur with an orderly distribution of genetic material to discrete locations within the cell. Its speed, its precision, and its importance to the life of the cell have all drawn attention to mitosis. Yet these are precisely the properties that probably account for our limited understanding of the process. Mitosis can take many forms, depending on the cells examined, in terms of the rate of separation, the number and distribution of microtubules, and the cellular location of the spindle, but it is probably most economical to begin with the assumption that the fundamental mechanism of mitosis is widespread, if not universal. Certainly the difficulties of analysis are great enough as it is, without invoking numerous idio-syncratic mechanisms for the process.

During mitosis, cells frequently undergo substantial cytoplasmic rearrangement. Cells in culture may round up, losing most attachments to the substrate and stress fibers. This loss of stress fibers does not appear to result from a change in vinculin phosphorylation [172]. Inter-mediate filaments may also rearrange extensively during mitosis. In some cells, vimentin filaments form a cage of filaments around the nucleus at mitosis, often accompanied by an increase in vimentin phosphorylation. The phosphorylation falls within 30 minutes of spreading [173, 174]. A different pattern of rearrangement is seen in cells that contain cytokeratins. Some rearrange the cytokeratins into slender protofilament threads and even into aggregates associated with the cytoskeleton. In telophase, the protein rearranges again into bundles. At least one cell type rearranges its vimentin filaments in this same way. Therefore the pattern of arrangement of intermediate filaments is under cellular control and is cell-type specific.

Microtubules undergo the greatest rearrangement in those cells which round up for mitosis. The extensive cytoplasmic arrays of microtubules disappear, and a microtubule-based spindle forms. The spindle segregates the chromosomes; its spatial and mechanical properties present a formidable challenge to biochemical explanations [175, 176, 177].

During prophase, the chromosomes condense and engage in saltatory motion in the region between the two centrioles and their associated material. The saltating chromosomes are accompanied by other particles engaging in gentle saltatory motion. Microtubules elongate from the poles with a uniform polarity; their slow assembly ends are associated with the poles and the fast assembly ends are distal [178, 179]. As chromosomes come into contact with microtubules, interaction occurs between the kinetochore and the microtubules. It is only when a chromosome has established a kinetochore association with microtubules from both poles that the saltatory motion ceases and the chromosome becomes immobile. When all chromosomes are anchored in this way, they organize into the metaphase plate in the equatorial plane of the spindle. At this time other particles in the region continue to saltate. Abruptly, when the sister chromatids separate, two forms of movement occur. (1) Chromosomes begin to approach the poles (anaphase A). (2) The spindle poles themselves move apart (anaphase B). There is no clear or simple relationship between these two aspects of anaphase in different species; their regulation may be independent. Anaphase motion continues until chromosomes reach the poles where they congregate and begin to reform as a nucleus during telophase.

Many aspects of this process present mechanistic challenges. The saltation of chromosomes during prophase cannot be due to microtubules, which have not yet reached them. Rather, the saltation seems to be mediated by the nuclear matrix. The stabilization of movement at metaphase strongly suggests that a balance of forces has been reached, in which the opposing poles exert an equal mechanical effect upon a single chromosome. Notably, if a metaphase chromosome is detached from one pole by laser irradiation, the chromosome immediately moves to the pole to which it is still attached. This ability for a single chromosome to anticipate anaphase, and the uniform polarity of microtubules in a half spindle, represent a minimum set of phenomena that must be accounted for by mechanisms of chromosome movement during

64

anaphase. The observations are relatively recent and present major difficulties to a number of the earlier theories of anaphase movement. The simplest model with which they are consistent is one recently proposed, in which the mechanochemical component during anaphase is the kinetochore, which is suggested to walk along the microtubule [180].

Several past models can be eliminated with greater or lesser degrees of certainty. No mitotic spindle is either cytochalasin sensitive or inhibited by antimyosin antibodies or similarly actin-specific poisons [175]. Similarly, all models that require antiparallel organization of microtubules in the spindle can now be more or less eliminated by the inappropriate orientation of microtubules.

Many aspects of mitosis are energy dependent. Metabolic inhibitors arrest the saltation of prometaphase. The spindle pole separation is clearly energy dependent, both *in vivo* and *in vitro*. The movement of chromosomes to the poles has been reported to be energy independent in a lysed cell system [181], but possibly under these conditions only an elastic component is observed since the motion is slower than that in the living cell. The role of non-microtubular elements in the spindle is only now coming to be appreciated. The smaller filamentous elements of the spindle have begun to be visualized in ways that suggest a role in spindle organization of chromosome movement (Fig. 4.2).

Moment-to-moment maintenance of cell shape is also dependent upon the assembly of cytoskeletal elements. Two experimental approaches are available to examine assembly of the interphase cytoskeleton, but both techniques have certain restrictions. One is microinjection of a fluorescently labeled protein that is assembly-competent *in vitro* [182–186]. The protein's location is followed by light microscopy; some kinetic properties can be measured by photobleaching. The structures visualized in this way are almost always faithful, i.e. the results agree with those of immunofluorescence. Microinjection studies suffer certain constraints. They can only identify structural elements in exchange with the soluble phase; elements sufficiently stable that little or no exchange occurs in the course of hours must be inaccessible. They cannot distinguish between assembly and exchange. The diffuse pattern immediately after injection changes to a discretely localized one; this is equally consistent with either assembly into structures undergoing exchange or with association with a restricted number of binding sites. Kinetics can be instructive. α Actinin and vinculin were both found to associate faithfully with adhesion plaques, but this required hours, while the structures with which they associate form and dissolve in minutes. This suggests that, although the proteins faithfully associated, the pathway by which they entered these structures was not the predominant one obtaining in the cell. Another constraint on these studies is that fluorescence exaggerates structure size. Salient points from such studies are these: (1) The assembly visualized by the process

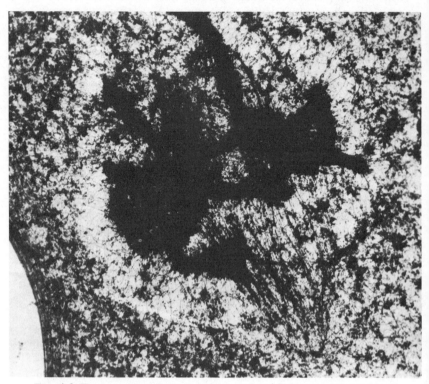

Fig. 4.2 Transmission electron micrograph of PtK cytoskeletal framework, showing nuclear matrix during mitosis.

is usually faithful. (2) It can be kinetically distinct from the appearance of the structures themselves. (3) In some cases, it is cell specific [184]. (4) The direction of assembly seen for tubulin was consistent with the fast assembly ends being distal. (5) The most mobile species of actin detected in any cell using this technique was larger than an actin monomer, suggesting that little if any actin in the cell is found as free G actin (186). These slower moving complexes could either be actin oligomers or actin associated with actin-specific proteins.

An alternate technique with a complementary constraint involves labeling cells *in vivo* with S^{35} methionine and studying the behavior of proteins either by autoradiography or biochemically. A large fraction of cytoskeletal proteins is found associated with the cytoskeleton near polyribosomes; these proteins rearrange with time, provided continued protein synthesis [65]. In cells large and symmetrical in shape, the pattern of rearrangement appears to be in a centripetal wave comparable to the slow component of axonal transport [22]. Using *in vitro* labeling of cytoskeletons, many of these proteins made puromycin-resistant associations before translation was complete [187]. Since the extractions used are several minutes long, those cytoskeletal proteins which exchange more frequently than once every several minutes would be lost from these structures.

4.6 The cytoskeleton and gene expression

The cytoskeleton is involved in gene expression in two ways. The cytoskeletal proteins themselves are regulated. In a broader role, the configuration of the cytoskeleton may affect gene expression either directly or with few intervening steps. All three major filament systems show developmental regulation of isoforms during embryogenesis and maturation of the organism. However, the mechanisms by which these changes in isoform are regulated are not yet known.

Short-term regulation of the cytoskeletal protein genes is accessible to experimentation in cells. Tubulin is probably in exchange with structures in the cell, and therefore a candidate for gene regulation by free monomer levels. Agents which raise the free monomer concentration decrease tubulin synthesis, and agents which sequester tubulin into polymers, either true microtubules or paracrystals, stimulate tubulin synthesis [188, 189]. Altered synthesis is accompanied by changed levels of mRNA, but measurements of gene activity in the nucleus show no difference [190].

Actin genes are differentially expressed depending on the shape of the cell. When 3T6 cells are suspended, protein synthesis is suppressed, but actin is specifically depressed relative to total protein synthesis. After replating, protein synthesis resumes. In the early period of spreading there is a relative over-synthesis of actin [191]. Changes in actin synthesis are also seen under chronic exposure to cytochalasin D. After a day, exposure to cytochalasins increases both the total actin in the cell and the relative rate of actin synthesis. At this drug level, cells are arborized in a different configuration from either the spread or suspended cells [192].

Some cellular control over cytoskeletal protein levels is exercised after protein synthesis. In muscle cells, the turnover of myofibril proteins is inversely related to the rate of contraction. Cells inhibited from contracting have a higher turnover rate of such myofibril proteins as α actinin, troponin C, muscle-specific myosin fast light chain, and $\alpha\beta$ tropomyosin. Non-contraction affects specifically myofibril proteins, since vimentin, desmin, and β and γ non-muscle actin are not affected. The increase in degradation leads to a decrease in myofibril proteins, accomplished without changing the rate of gene expression [193]. In contrast, many times the isoform expressed by a muscle is directly affected by innervation; the gene transcribed changes with different sources of innervation. At least some intermediate filament proteins are also affected by cell configuration. In suspended cells, the relative

67

synthesis of vimentin is almost obliterated but returns to normal when the cells are replated [194].

Two filament systems can be affected at the same time, either in parallel or in complementary fashion. Pre-adipocytes convert from a fibroblast-like shape to a nearly spherical cell. This shape change accompanies large decreases in the synthesis of β and γ actin, vimentin, and α and β tubulin. The decreased synthesis results from lower levels of translatable RNA. The changes in synthesis occur very early in differentiation, and may influence the later steps of adipocyte differentiation. Co-ordinated changes in cytoskeletal gene expressions are not restricted to Metazoa, as the amoeboflagellate *Naegleria* undergoes a reversible change from an amorphous walking amoeba to a streamlined biflagellated cell. This transformation involves a decrease in actin synthesis and an increase in tubulin synthesis and of other proteins required for the axoneme [118].

The cytoskeleton itself plays either a direct or a proximal role in regulating several levels of gene expression. Both microfilaments and microtubules can modulate a cell's response to signals that initiate DNA synthesis in 3T3 cells. The cell's ability to respond to serum and five other growth factors after deprivation is blocked reversibly by low doses of dihydrocytochalasin B [195]. The drug does not reduce either glucose or thymidine transport so that the inhibition of DNA synthesis cannot be simple nutritional starvation. Finally this microfilament-dependent step is not required in transformed cells, since two different transforming agents relieve the sensitivity to cytochalasin.

Alterations of microtubules can also alter growth rates as a function of the state of the cell. At low density, cells are inhibited by microtubule depolymerizing agents, while cells at high density show a stimulation of DNA synthesis. There is cell specificity involved, since established 3T3 cells show less inhibition when sparse than primary embryonic cells. The stimulation of DNA synthesis seen in confluent cultures is antagonized by taxol, and taxol alone can prevent the mitogenic effects of a broad spectrum of stimuli [196].

In addition to these specific filament effects, there is a large body of alterations in cell expression with changes of cell shape. For example, the mitotic rate of cells in culture is closely correlated to their height. In a confluent monolayer where the cells are substantially thicker than those in a sparse culture, the tall cells are also less responsive to stimulation by serum and other growth factors. Parallel changes in growth rates and serum sensitivity can be generated by changing the cell height [197].

In other cell lines, direct modulation of cell shape by changing the substrate adhesiveness alters gene expression. For example, chondrocytes sythesize specific products in certain configurations, although the rate of total protein synthesis is not affected [197]. Similarly, pre-adipocytes differentiate when a round shape is imposed upon them.

Perhaps more striking are the consequences of withdrawal of attachment, discussed in connection with transformation (*v.i.*). The complementary roles of cell shape and the extracellular material in cell expression are visible in mammary glands. These cells grow in contact with or inside of collagen gels, their normal extracellular material. However, their shape on the collagen correlates with the fidelity of their differentiated expression. Cells grown flat showed no morphological differentiation and did not accumulate milk protein. On floating gels, two cell types were visible; a cuboidal one in contact with the medium responded to prolactin by lipid synthesis, while the other contained many microfilaments and resembled the myoepithelial cells of the mammary gland. Cells actually embedded in gels formed hollow duct-like structures [198].

The full range of interaction between the cytoskeleton and gene expression is perhaps clearest in the contrast between transformed and normal cells (*v.i.*). However, from cases discussed so far, it is clear that a network of interactions is continuously operating in all cells. The cytoskeleton controls the display of, and in some cases the responsiveness of, the membrane proteins, including receptors. Occupancy of membrane receptors by soluble factors and interaction of membrane proteins with the solid extracellular matrix can affect cytoskeletal organization. Thus there is potential for a feedback between the cytoskeleton and the exterior space, one that is independent of changes in gene expression. Strikingly, the extracellular matrix plays a major role in the expression of differentiated phenotypes. As we have seen, some of this control seems to be modulated directly by cytoskeletal configuration itself. The role of the extracellular matrix may be to alter the state of the cytoskeleton, so that the alterations of the cytoskeleton can directly or proximally alter gene expression.

4.7 Transformation

Although transformation is a complex process, some generalizations can be made. In benign growths, cellular architecture is slightly affected or not at all, and the cell's relationship to the ECM is largely unchanged. In malignant growths, the cellular architecture is greatly disturbed, and the relationship between the malignant cell and its tissue of origin is frequently difficult to discern. This change in architecture may be a change in organization, since it is known that tumor cells generally preserve the intermediate filament proteins of their tissue of origin. In addition, malignant cells are invasive and violate the ECM that surrounds them, showing that the matrix no longer constrains their behavior as it does normal cells. The final stage of malignancy is metastasis, in which cells escape from the tumor, disperse through the body and establish a colony in other tissues.

Numerous correlates of these behaviors can be observed in cultured cells. Transformation in culture has been measured as the production of plasminogen activator, as a reduction in serum requirement, or as a loss of contact inhibition of growth and movement. These two properties are also called density-dependent inhibition of growth or movement. Normal cells respond to contacts with other cells by stopping locomotion and reversing direction. If the immediate vicinity of the cell is occupied by so many cells that reversal cannot produce an unobstructed path, the cell may acquire a more rounded shape. At such cell densities overall growth is frequently inhibited for normal cells. Transformed cells do not retract when making contact with other cells, but will rather overlap them and crawl either over or under them without changing direction. Transformed cells are also less sensitive to the immediate density of cells around them for growth. The single strongest correlate of tumorigenesis in the body is the loss of anchorage dependence *in vitro*; those cells which will grow suspended in liquid medium almost uniformly are those which will generate tumors in animals. These various properties are not expressed in unison by all transformed cells. The commonest sequence is a reduction in the requirement for serum, followed by a loss of requirement for attachment, but appropriate selection can generate cell lines which require serum but not attachment [199].

Just as, in the body, the transition from benign to malignant is accompanied by gross changes in cellular architecture, so also in culture can cell series be found in which the architecture is progressively more disarranged, while the line expresses a more transformed biochemical phenotype. In the series from normal mouse embryonic fibroblasts through virally transformed 3T3 cells, cells range from having many cables and a well-spread shape to being rounder, less occupied by stress fibers and less orderly in their relationships to each other. Within this series, the greater the degree of transformation, the more biochemical processes are resistant to the loss of a substrate. Thus, when a normal embryonic fibroblast is suspended, DNA, RNA and protein synthesis all shut down promptly. At intermediate stages in the series, some events continue, such as hnRNA synthesis, but message production stops, while the virally transformed cell is capable of synthesizing proteins and undergoing division in suspension [200]. A cell in which such processes have been shut down by suspension will re-express protein synthesis with attachment, but will not begin RNA or DNA synthesis until the cell is allowed to spread; the cell distinguishes between attachment and overall cell configuration [201].

Loss of stress fibers and cellular organization has been widely observed under the influence of many transforming agents. Purified src kinase microinjected into the cell eliminates stress fibers within 30 minutes, which suggests an early role for cytoskeletal disorganization [202]. Similarly, individuals predisposed to certain forms of cancer show altered organization of stress fibers before the presence of cancer, as well as lower levels of stress fibers in individuals with those forms of cancers [203]. Both observations point to an early and possibly primary role in cytoskeletal disorganization in the process of transformation.

Phenocopies of the cytoskeletal organization can be induced by other agents as well. Tumor promoters such as TPA or proteases with the same specificity as plasmin can both generate a comparably disorganized cytoskeleton. However, imposing the change in organization does not lead to a stably transformed cell [204]. This further suggests that although cytoskeletal disorganization is a frequent correlate of transformation, it is not by itself enough to generate the stably transformed state.

Certain surface properties of transformed cells suggest a cytoskeletal role in transformation. Many transformed cells display less fibronectin on their surfaces, possibly because of secretion of plasminogen activator by transformed cells. Similarly, many transformed cells show resistance to the con-A modulation of normal cells. Since fibronectin is essential for migration and mediates many ECM events, this suggests that one site for disorganization in transformed cells is at the interface between the cytoskeleton and the plasma membrane.

These observations taken together lead to the following generalizations. A change in cytoskeletal organization generally parallels an increase in transformation; in many cases, the change in cytoskeletal organization can be detected before the tumorigenic state. However, imposing a phenocopy of the cytoskeletal organization, even by tumor promoters, is not sufficient to generate the transformed state, suggesting that a second lesion in cellular regulation is required for full transformation. Finally, in those transformed cells which show disorganized cytoskeletal arrangements, if a more normal morphology is imposed upon them by, for example, exogenous fibronectin, the cytoarchitecture can be normalized and cell behavior is frequently more normal.

The actin-based networks, particularly the stress fibers, are the cytoskeletal structures for which the most correlative data is available in normal and transformed cells. Microtubules may also be disorganized by transformation, but this is difficult to resolve. Anything that prevents cell spreading will interfere with the extension of microtubule networks. Also, in a rounded cell, immunofluorescence patterns are difficult to analyse. Within these constraints, it has been suggested that transformed cells do contain microtubule networks but ones which necessarily are more rounded than in a fully spread cell.

Intermediate filament involvement is suggested by two further observations. Normal cells treated with cytochalasin arborize, i.e. extend long slender processes from a rounded central area. These arborized processes contain many intermediate filaments. Normal cells treated with transforming viruses cannot form such extended processes [205]. Since cytochalasin prevented filopodial extension and ruffling, these processes must extend by some other interaction with a substrate, which presumably involves the associations that have been seen between intermediate filaments and the cell membrane. A second striking observation is that the imposed cell shape can change the ability of transformed cells to metastasize. Melanoma cells kept spherical are dramatically more metastatic. One day in culture under spreading conditions reverses this increase in metastasis, which is therefore not likely to be merely a selective process [194]. The absence of any drug specific for intermediate filaments makes it difficult to test whether the two are directly related, but these observations suggest that intermediate filament organization may also play a role in the expression of transformation.

Normal cells are characterized by both a carefully regulated, organized and specific cytoskeleton and by an appropriate interaction with the exterior, particularly with the ECM and other cells. Both this architecture and social behavior appear to be mediated through the cytoskeleton; both are defective in transformed cells. The direction of causation and the mechanisms involved are slowly being understood.

Much is known about the protein chemistry of cytoskeletal proteins. As potential directions of interaction are appreciated, more proteins are discovered and their properties under physiological and non-physiological conditions become clearer. However, what is presently known about cytoskeletal proteins is not enough to account for cytoskeletal architecture.

Cytoskeletal architecture varies with cell type, differentiation and cell growth. In most cells, cytoskeletal organization is anisotropic and asymmetric. Most cells can undergo continuous, dynamic reorganization of cytoskeletal elements. A few cytoskeletal structures, such as the centriole, are reasonable candidates for spontaneous assembly from solution, but most appear to be in a state of continuous rearrangement. The highly localized behavior of cytoskeletal proteins, the tissue specific isoforms and their functions, and the carefully regulated rearrangement of these proteins are all aspects of cytoskeletal architecture that need to be explained by protein chemistry but which are not yet understood.

Finally, although the location and rearrangement of these proteins within the cell has been documented for many cell types, our knowledge of cytoskeletal architecture does not yet account for what might be called cellular choreography, the controlled rearrangement in time and space of most of the visible components of the cell. It is this behavior that allows the cell to maintain itself, to grow, and to divide. It is choreography that is the phenomenon to be explained; questions posed about architecture and protein chemistry are most illuminating when they are framed to explain this behavior.

References

Particularly helpful sources are marked so:*

 [1] DeRosier, D.J. *et al.* (1980), *Nature,* **287**, 291–296.
*[2] Clarke, M. and Spudich, J.A. (1977), *Ann. Rev. Biochem.,* **46**, 797–822.
 [3] Osborn, M. and Weber, K. (1982), *Cell,* **31**, 303–306.
 [4] Geisler, N. and Weber, K. (1982), *EMBO J.,* **1**, 1649–1656.
*[5] Korn, E.D. (1982), *Physiological Rev.,* **62**, 672–737.
*[6] Schliwa, M. (1981), *Cell,* **25**, 587–590.
 [7] Southwick, F.S. and Hartwig, J.H. (1982), *Nature,* **297**, 303–307.
 [8] Yamamoto, K. *et al.* (1982), *J. Cell Biol.,* **95**, 711–719.
 [9] Glenney, J.R. *et al.* (1981), *J. Biol. Chem.,* **256**, 9283–9289.
[10] Repasky, E.A. *et al.* (1982), *Cell,* **29**, 821–833.
[11] Glenney, J.R., Jr. *et al.* (1982), *Proc. Natl. Acad. Sci. USA,* **79**, 4002–4005.
[12] Levine, J. and Willard, M. (1981), *J. Cell Biol.,* **90**, 631–643.
*[13] Timasheff, S.N. and Grisham, L.M. (1980), *Ann. Rev. Biochem.,* **49**, 565–591.
[14] Wegner, A. (1976), *J. Mol. Biol.,* **131**, 839–853.
[15] Sherline, P. and Schiavone, K. (1977), *Science* (Wash. D.C.), **198**, 1038–1040.
[16] Weingarten, M.D. *et al.* (1975), *Proc. Natl. Acad. Sci. USA,* **72**, 1858–1862.
[17] Cleveland, D.W. *et al.* (1977), *J. Mol. Biol.,* **116**, 207–225.
[18] Suprenant, K.A. and Dentler, W.L. (1982), *J. Cell Biol.,* **93**, 164–174.
[19] Pallas, D. and Solomon, F. (1982), *Cell,* **30**, 407–414.
[20] Theurkauf, W.E. and Vallee, R.B. (1982), *J. Biol. Chem.,* **257**, 3284–3290.
[21] Vallee, R.B. and Davis, S.E. (1983), *Proc. Natl. Acad. Sci. USA,* **80**, 1342–1346.
*[22] The Cytoplasmic Matrix and the Integration of Cellular Function, *J. Cell Biol.,* in press.
[23] Geisler, N. and Weber, K. (1981), *J. Mol. Biol.,* **151**, 565–571.
[24] Steinert, P.M. (1978), *J. Mol. Biol.,* **123**, 49–70.
[25] Geisler, N. and Weber, K. (1982), *EMBO. J.,* **1**, 1649–1656.
[26] Osborn, M. and Weber, K. (1982), *Cell,* **31**, 303–306.
[27] Steinert, P.M. *et al.* (1981), *Proc. Natl. Acad. Sci. USA,* **78**, 3692–3696.

[28] Geisler, N. and Weber, K. (1981), *Proc. Natl. Acad. Sci. USA,* **78**, 4120–4123.

[29] Walter, M.F. *et al.* (1983), *J. Cell Biol.,* **97**, 223a.

[30] Nelson, W.J. and Traub, P. (1981), *FEBS,* **161**, 51–57.

[31] Day, W.A. (1980), *J. Ultrastruct. Res.,* **70**, 1–7.

[32] Azanza, J.-L. *et al.* (1979), *Biochem. J.,* **183**, 339–347.

[33] Wiche, G. *et al.* (1983), *J. Cell Biol.,* **97**, 887–901.

[34] Granger, B.L. *et al.* (1982), *J. Cell Biol,* **92**, 299–312.

[35] Breckler, J. and Lazarides, E. (1982), *J. Cell Biol.,* **92**, 795–806.

[36] Bloom, G.S. and Vallee, R.B. (1983), *J. Cell Biol.,* **96**, 1523–1531.

[37] Davis, J. and Bennett, V.J. (1982), *J. Biol. Chem.,* **257**, 5816–5820.

[38] Pfeffer, S.R. *et al.* (1983), *J. Cell Biol,* **97**, 40–47.

*[39] Amos, W.B. (1975), in *Molecules and Cell Movement* (eds S. Inoué and R. E. Stephens), Raven Press, New York.

[40] McKeithan, T.W. *et al.* (1983), *J. Cell Biol.,* **96**, 1056–1063.

[41] Sefton, B.M. *et al.* (1981), *Cell,* **24**, 165–174.

[42] Adelstein, R.S. and Klee, C.B. (1980), in *Calcium and Cell Function*, Academic Press, New York.

[43] Spudich, J.A. *et al.* (1982), *CSH Symp. Quant. Biol.,* **46**, 553–561.

[44] Craig, S.W. and Ko, C.G. (1982), *Fed. Proc.,* **41**, 1421.

[45] Tilney, L.G. and Detmers, P. (1975), *J. Cell Biol.,* **66**, 508–520.

[46] Pinder, J.C. and Gratzer, W.B. (1983), *J. Cell Biol.,* **96**, 768–775.

[47] Branton, D. *et al.* (1981), *Cell,* **24**, 24–32.

[48] Nakashima, D. and Beutler, E. (1979), *Proc. Natl. Acad. Sci. USA,* **76**, 935–938.

[49] Gillies, R.J. (1982), *Trends in Biochem. Sci.,* **7**, Feb, 41–42.

[50] Blikstad, I. and Lazarides, E. (1983), *J. Cell Biol.,* **96**, 1803–1808.

[51] Bartelt, D.C. *et al.* (1982), *J.Cell Biol.,* **95**, 278–284.

[52] Granger, B.L. and Lazarides, E. (1982), *Cell,* **30**, 263–275.

[53] Blikstad, I. and Lazarides, E. (1983), *J. Cell Biol.,* **96**, 1803–

*[54] Carroll, R.C. and Cox, A.C. (1983), *Surv. Synth. Path. Res.,* **2**, 21–33.

[55] Gonnella, P.A. and Nachmias, V.T. (1981), *J. Cell Biol.,* **89**, 146–151.

*[56] Debus, E. *et al.* (1981), *Eur. J. Cell Biol.,* **24**, 45–52.

[57] Lazarides, E. (1976), *Cell Motility,* **3**, CSH Conf. on Cell Proliferation, 347–360.

[58] Heggeness, M.H. *et al.* (1977), *Proc. Natl. Acad. Sci. USA,* **74**, 3883–3887.

[59] Bretscher, A. and Weber, K. (1980), *J. Cell Biol.,* **86**, 335–340.

[60] Avnur, Z. *et al.* (1983), *J. Cell Biol.,* **96**, 1622–1630.

[61] Weber, K. *et al.* (1975), *Proc. Natl. Acad. Sci. USA,* **72**, 459–463.

[62] Henderson, D. and Weber, K. (1980), *Exp. Cell Res.,* **129**, 441–453.

[63] Ip, W. *et al.* (1983), *J. Cell Biol.,* **96**, 401−408.
[64] Schliwa, M. and van Blerkom, J. (1981), *J. Cell Biol.,* **90**, 222−235.
[65] Fulton, A.B. *et al.* (1980), *Cell,* **20**, 849−857.
[66] Travo, P. *et al.* (1982), *Exp. Cell Res.,* **139**, 87−94.
[67] Sobieszek, A. (1977), in *Biochemistry of Smooth Muscle,* 413−443.
[68] Osborn, M. *et al.* (1981), *Different.,* **20**, 196−202.
[69] Traeger, L. and Goldstein, M.A. (1983), *J. Cell Biol.,* **96**, 100−103.
[70] Pardo, J.V. *et al.* (1983), *J. Cell Biol.,* **97**, 1081−1088.
[71] Nelson, W.J. and Lazarides, E. (1983), *Proc. Natl. Acad. Sci. USA,* **80**, 363−367.
[72] Tokuyasu, K.T. *et. al.* (1983), *J. Cell Biol.,* **96**, 1736−1742.
[73] Pardo, J.V. *et al.* (1983), *Proc. Natl. Acad. Sci. USA,* **80**, 1008−1012.
[74] Craig, S.W. and Pardo, J.V. (1983), *Cell Motility,* **3**, 449−462.
[75] Tokuyasu, K.T. *et al.* (1983), *J. Cell Biol.,* **96**, 1727−1735.
[76] Lazarides, E. *et al.* (1982), *CSH Symp. Quant. Biol.,* **46**, 351−378.
[77] Wang, K. and Williamson, C. L. (1980), *Proc. Natl. Acad. Sci. USA,* **77**, 3254.
[78] Reinach, F.C. *et al.* (1983), *J. Cell Biol.,* **96**, 297−300.
[79] Wallimann, T. *et al.* (1983), *J. Cell Biol.,* **96**, 1772−1779.
[80] Pardo, J.V. *et al.* (1983), *Cell,* **32**, 1093−1103.
[81] Gomer, R.H. and Lazarides, E. (1981), *Cell,* **23**, 524−532.
[82] Fallon, J.R. and Nachmias, V.T. (1980), *J. Cell Biol.,* **87**, 237−247.
[83] Bader, D. *et al.* (1982), *J. Cell Biol.,* **95**, 763−770.
[84] Gard, D.L. and Lazarides, E. (1980), *Cell,* **19**, 263−275.
[85] Bennett, G.S. *et al.* (1978), *Different.,* **12**, 71−82.
[86] Osborn, M. *et al.* (1982), *CSH Symp. Quant. Biol.,* **46**, 413−429.
[87] Wehland, J. *et al.* (1979), *J. Cell Sci.,* **37**, 257−273.
[88] Sanger, J.W. *et al.* (1983), *J. Cell Biol.,* **96**, 961−969.
[89] Herman, I.M. and Pollard, T.D. (1981), *J. Cell Biol.,* **88**, 346−351.
[90] Herman, I.M. *et al.* (1981), *J. Cell Biol.,* **90**, 84−91.
[91] Frixione, E. (1983), *J. Cell Biol.,* **96**, 1258−1265.
[92] Euteneuer, U. and McIntosh, J.R. (1981), *Proc. Natl. Acad. Sci. USA,* **78**, 372−376.
[93] Franke, W.W. *et al.* (1979), *Different.,* **15**, 7−15.
[94] Tseng, S.C.G. *et al.* (1982), *Cell,* **30**, 361−372.
[95] Moll, R. *et al.* (1982), *J. Cell Biol.,* **95**, 285−295.
[96] Schmid, E. *et al.* (1983), *J. Cell Biol.,* **96**, 37−50.
[97] Henderson, D. and Weber, K. (1981), *Exp. Cell Res.,* **132**, 297−311.
[98] Mooseker, M.S. (1983), *Cell,* **35**, 11−13.
[99] Tilney, L.G. *et al.* (1980), *J. Cell Biol.,* **86**, 244−259.
[100] Nagle, B.W. and Burnside, B. (1984), *J. Cell Biol.,* in press.
[101] Owaribe, K. and Masuda, H. (1982), *J. Cell Biol.,* **95**, 310−315.

[102] Hubbard, A.L. and Ma, A. (1983), *J. Cell Biol.*, **96**, 230–239.
[103] Porter, K.R. and Anderson, K.L. (1982), *Eur. J. Cell Biol.*, **29**, 83–96.
[104] Schliwa, M. *et al.* (1981), *Proc. Natl. Acad. Sci. USA*, **78**, 4329–4333.
[105] Maupin, P. and Pollard, T.D. (1983), *J. Cell Biol.*, **96**, 51–62.
[106] Jahn, W. (1980). *Eur. J. Cell Biol.*, **20**, 301–304.
[107] Gabbiani, G. *et al.* (1983), *Proc. Natl. Acad. Sci. USA*, **80**, 2361–2364.
[108] Selden, S.C. III *et al.* (1981), *J. Cell Physiol.*, **108**, 195–211.
[109] Ausprunk, D.H. and Berman, H.J. (1978), *Tissue and Cell*, **10**, 707–724.
[110] Pollack, R. *et al.* (1975), *Proc. Natl. Sci. USA*, **72**, 994–998.
[111] Edelman, G. and Yahara, I. (1976), *Proc. Natl. Sci. USA*, **73**, 2047–2051.
[112] Leavitt, J. *et al.* (1982), *Cell*, **28**, 259–268.
[113] Witt, D.P. *et al.* (1983), *J. Cell Biol.*, **96**, 1766–1771.
[114] Shriver, K. and Rohrschneider, L. (1981), *J. Cell Biol.*, **89**, 525–535.
[115] Osborn, M. and Weber, K. (1983), *Lab. Invest.*, **48**, 372–394.
[116] Travis, J.L. and Allen, R.D. (1981), *J. Cell Biol.*, **90**, 211–221.
[117] Nagai, R. *et al.* (1978), *J. Cell Sci.*, **33**, 205–225.
[118] Fulton, C. and Lai, E.Y. (1982), in *Microtubules in Microorganisms* (eds P. Cappucinelli and N. R. Morris) 235–256.
[119] Huang, B. *et al.* (1981), *J. Cell Biol.*, **88**, 80–88.
[120] Minami, S.A. *et al.* (1981), *Cell*, **24**, 89–95.
[121] Soll, D.R. (1984), in *The Microbial Cell Cycle* (eds T. Nurse and E. Strieblova), CRC Press Inc., Boca Raton, FL., in press.
*[122] Lloyd, C.W. (1982), *The Plant Cytoskeleton*, Academic Press, New York.
[123] Wick, S.M. *et al.* (1981), *J. Cell Biol.*, **89**, 685–690.
[124] Hardham, A.R. *et al.* (1980), *Planta (Berl.)*, **149**, 181–195.
[125] Wick, S.M. and Duniec, J. (1983), *J. Cell Biol.*, **97**, 235–243.
*[126] Dustin, P. (1978), *Microtubules*, Springer-Verlag, Berlin.
[127] Horwitz, S.B. *et al.* (1982), *CSH Symp. Quant. Biol*, **46**, 219–226.
[128] Flanagan, M.D. and Lin, S. (1980), *J. Biol. Chem.*, **255**, 835–838.
[129] Atlas, S.J. and Lin, S. (1978), *J. Cell Biol.*, **76**, 360–370.
[130] Schliwa, M. (1982), *J. Cell Biol.*, **92**, 79–91.
[131] Bershadsky, A.D. *et al.* (1980), *Exp Cell Res.*, **127**, 421–429.
[132] Sharpe, A.H. *et al.* (1980), *Proc. Natl. Acad. Sci. USA*, **77**, 7267–7271.
[133] Lazarides, E. and Gard, D.L. (1982), in *Gene Regulation* (ed. B.W. O'Malley) Academic Press, New York, 343–358.
[134] Kruh, J. (1982), *Mol. Cell. Biochem.*, **42**, 65–82.
[135] Huang, B. *et al.* (1982), *Cell*, **28**, 115–124.
[136] Allen, R.D. (1960). *J. Biophys. Biochem. Cytol.*, **8**, 379–396.
[137] Allen, R.D. and Taylor, D.L. (1975), in *Molecules and Cell Movement* (eds S. Inoué and R.E. Stephens), Raven Press, New York, pp. 351–378.

[138] Bhisey, A.N. and Freed, J.J. (1971), *Exp. Cell Res.,* **64**, 419.

[139] Cheung, H.T. *et al.* (1978), *Exp. Cell Res.,* **111**, 95–103.

[140] Malawista, S.E. and Chevance, A.DeB. (1982). *J. Cell Biol.,* **95**, 960–973.

[141] Zigmond, S.H. *et al.* (1981), *J. Cell Biol.,* **89**, 585–592.

[142] Goldman, R.D. (1971), *J. Cell Biol.,* **51**, 752–762.

[143] Albrecht-Buehler, G. and Lancaster, R.M. (1976), *J. Cell Biol.,* **71**, 370–382.

[144] Gawlitta, W. *et al.* (1981), *Eur. J. Cell Biol.,* **26**, 83–90.

[145] Gotlieb, A.I. *et al.* (1983), *J. Cell Biol.,* **96**, 1266–1272.

[146] Albrecht-Buehler, G. (1977), *Cell,* **12**, 333–339.

[147] Albrecht-Buehler, G. (1977), *Cell,* **11**, 395–404.

[148] Ali, I.U. and Hynes, R.O. (1978), *Cell,* **14**, 439–446.

[149] Moore, L. and Pastan, I. (1979), *J. Cell Physiol.,* **101**, 101–108.

[150] Goldman, R.D. *et al.* (1973), *Proc. Natl. Acad. Sci. USA,* **70**, 750–754.

[151] Steinberg, M.S. and Poole, T.J. (1982), in *Developmental Order: Its Origin and Regulation,* Alan R. Liss, Inc., New York, pp. 351–378.

[152] Solomon, F. and Magendantz, M. (1981), *J. Cell Biol.,* **89**, 157–161.

[153] Carroll, R.C. *et al.* (1982), *Cell,* **30**, 385–393.

[154] Webb, W.W. *et al.* (1982), *Biochem. Soc. Symp.,* **46**, 191–205.

[155] Prives, J. *et al.* (1982), *J. Cell Biol.,* **92**, 231–236.

[156] Geiger, B. *et al.* (1982), *J. Cell Biol.,* **95**, 137–143.

[157] Edelman, G.M. (1976), *Science,* **192**, 218–226.

[158] Bourguignon, L.Y.W. *et al.* (1982), *J. Cell Biol.,* **95**, 793–797.

*[159] Oliver, J.M. and Berlin, R.D. (1982), *Intl. Rev. Cytol.,* **74**, 55–94.

*[160] Bissell, M.J. *et al.* (1982), *J. Theor. Biol.,* **99**, 31–68.

[161] Allen, R.D. *et al.* (1981) *Cell Motility,* **1**, 291–302.

*[162] Buckley, I.K. (1974), *Tissue and Cell,* **6**, 1–20.

[163] Buckley, I. and Steward, M., *Cell Motility,* in press.

[164] Forman, D.S. *et al.* (1983), *J. Neurosci.,* **3**, 1279–1288.

[165] Berlinrood, M. *et al.* (1972), *J. Cell Sci.,* **11**, 875–886.

[166] Bajer, A. and Allen, R.D. (1966), *J. Cell Sci.,* **1**, 455–462.

[167] Schliwa, M. (1982), *Methods in Cell Biology,* **25**, 285–312.

[168] Brady, S.T. *et al.* (1982), *Science,* **218**, 1129–1131.

[169] Busson-Mabillot, S. *et al.* (1982), *J. Cell Biol.,* **95**, 105–117.

[170] Tokunaka, S. *et al.* (1983), *Different.,* **24**, 39–47.

[171] Tanenbaum, S.W. (1978), *Cytochalasins: Biochemical and Cell Biological Aspects,* North-Holland Publishing Co., Amsterdam.

[172] Rosok, M.J. and Rohrschneider, L.R. (1983), *Mol. Cell Biol.,* **3**, 475–479.

[173] Franke, W. W. *et al.* (1982), *Cell,* **30**, 103–113.

[174] Evans, R.M. and Fink, L.M. (1982), *Cell,* **29**, 43–52.

*[175] Pickett-Heaps, J.D. *et al.* (1982), *Cell,* **29**, 729–744.

[176] McIntosh, J.R. (1982), in *Developmental Order: Its Origin and Regulation,* (ed. S. Subtelny) Alan R. Liss, Inc., New York, 77–115.

[177] Inoué, S. and Ritter, H. (1975), *Molecules and Cell Movement*, Raven Press, New York.
[178] Telzer, R.B. and Haimo, L.T. (1981), *J. Cell Biol.*, **89**, 373–378.
[179] Euteneuer, U. *et al.* (1983), *J. Cell Biol.*, **97**, 202–208.
[180] Pickett-Heaps, J.D. and Spurck, T.P. (1982), *Eur. J. Cell Biol.*, **28**, 77–82.
[181] Cande, W.Z. (1983), *Nature*, **304**, 557–558.
[182] Glacy, S.D. (1983), *J. Cell Biol.*, **96**, 1164–1167.
[183] Feramisco, J.R. (1979), *Proc. Natl. Acad. Sci. USA*, **76**, 3967–3971.
[184] Keith, C.H. *et al.* (1981), *J. Cell Biol.*, **88**, 234–240.
[185] Burridge, K. and Feramisco, J.R. (1980), *Cell*, **19**, 587–595.
[186] Kreis, T.E. *et al.* (1982), *Cell*, **29**, 835–845.
[187] Fulton, A.B. and Wan, K.M. (1983), *Cell*, **32**, 619–625.
[188] Ben-Ze'ev, A. *et al.* (1979), *Cell*, **17**, 319–325.
[189] Cleveland, D.W. *et al.* (1981), *Cell*, **25**, 537–546.
[190] Cleveland, D.W. and Havercroft, J.C. (1983), *J. Cell Biol.*, **97**, 919–924.
[191] Farmer, S. R. *et al.* (1983), *Mol. Cell. Biol.*, **3**, 182–189.
[192] Tannenbaum, J. and Godman, G.C. (1982), *Mol. Cell Biol.*, **3**, 132–142.
[193] Crisona, N.J. and Strohman, R.C. (1983), *J. Cell Biol.*, **96**, 684–692.
[194] Raz, A. and Ben-Ze'ev, A. (1983), *Science*, **221**, 1307–1310.
[195] Maness, P.F. and Walsh, R.C., Jr. (1982), *Cell*, **30**, 253–262.
[196] Wang, Z.-W. and Rozengurt, E. (1983), *J. Cell Biol.*, **96**, 1743–1750.
*[197] Folkman, J. and Moscona, A. (1978), *Nature*, **273**, 345–349.
[198] Haeuptle, M.-T. *et al.* (1983), *J. Cell Biol.*, **96**, 1425–1434.
[199] Powers, S. *et al.* (1982), *CSH Conference on Cell Proliferation*, **9**, 243–258.
[200] Wittelsberger, S.C. *et al.* (1981), *Cell*, **24**, 859–866.
[201] Ben-Ze'ev, A. *et al.* (1980), *Cell*, **21**, 365–372.
[202] Maness, P.F. and Levy, B.T. (1983), *Mol. Cell. Biol.*, **3**, 102–112.
[203] Nicholson, N.B. *et al. International Cell Biol. 1980–1981*, Springer-Verlag, Berlin, 331–335.
[204] Rifkin, D.B. *et al.* (1979), *Cell*, **18**, 361–368.
[205] Menko, A.S. *et al.* (1983), *Mol. Cell. Biol.*, **3**, 113–125.

Index